PRACTICAL SYNTHESIS OF HIGH-PERFORMANCE ANALOG CIRCUITS

PRACTICAL SYNTHESIS OF HIGH-PERFORMANCE ANALOG CIRCUITS

by

Emil S. Ochotta
Xilinx, San Jose

Tamal Mukherjee
Carnegie Mellon University

Rob A. Rutenbar
Carnegie Mellon University

L. Richard Carley
Carnegie Mellon University

KLUWER ACADEMIC PUBLISHERS
Boston / Dordrecht / London

Distributors for North, Central and South America:
Kluwer Academic Publishers
101 Philip Drive
Assinippi Park
Norwell, Massachusetts 02061 USA
Telephone (781) 871-6600
Fax (781) 871-6528
E-Mail <kluwer@wkap.com>

Distributors for all other countries:
Kluwer Academic Publishers Group
Distribution Centre
Post Office Box 322
3300 AH Dordrecht, THE NETHERLANDS
Telephone 31 78 6392 392
Fax 31 78 6546 474
E-Mail <services@wkap.nl>

Electronic Services <http://www.wkap.nl>

Library of Congress Cataloging-in-Publication Data

A C.I.P. Catalogue record for this book is available
from the Library of Congress.

Printed on acid-free paper.

Printed in the United States of America

For Irene, Nicholas, and for Bin - ESO

Contents

List of Figures

List of Tables

Preface

Market competition and the astounding pace of technological innovation exert tremendous pressure on circuit design engineers to turn ideas into products quickly and get them to market. In digital Application Specific Integrated Circuit (ASIC) design, computer aided design (CAD) tools have substantially eased this pressure by automating many of the laborious steps in the design process, thereby allowing the designer to leverage his design expertise. For example, general purpose digital synthesis techniques that can reliably transform a behavioral circuit description into a collection of gates allow designers to focus on the more abstract and understandable behavioral description of the design.

But the world is not solely digital.

Cellular telephones, magnetic disk drives, neural networks and speech recognition systems are a few of the recent technological innovations that rely on a core of analog circuitry and exploit the density and performance of mixed analog/digital ASICs. To maximize profit, these mixed-signal ASICs must also make it to market as quickly as possible. However, although the engineer working on the digital portion of the ASIC can rely on sophisticated CAD tools to automate much of the design process, there is little help for the engineer working on the analog portion of the chip. With the exception of simulators to verify the circuit design when it is complete, there are almost no general purpose CAD tools that an analog design engineer can leverage to automate the analog design flow and reduce his time to market.

This does not mean that no one has attempted to create general purpose analog CAD tools. Although there is little discussion of synthesis of analog circuits from a behavioral model, there has been significant effort in the area of synthesis for performance. This kind of synthesis assumes a known qualitative behavior, such as amplification, and seeks to design a circuit that meets the many performance specifications that quantify that behavior. Unlike digital design, a set of 30 or more different performance specifications is not unusual for a single circuit, and most of these are difficult to evaluate even when given the circuit solution, making for a daunting synthesis task. Despite the best efforts of researchers, to industrial analog designers most of the analog synthesis tools published to date have been little more than a curiosity. These tools typically encode circuit-specific performance evaluation code or even a customized design strategy. As a result, these tools work for only a predetermined handful of circuits and over a limited range of performance. Moreover, the circuits produced are not comparable to those produced by the manual design process, often suffering from poor performance or the inability to tolerate the inevitable variations in manufacturing process or operating environment. Because of these limitations, these analog synthesis tools are poor cousins to their digital counterparts. Analog designers continue to wait for an analog synthesis tool that is practical: a tool that can automate part of the design process, leveraging the efforts of an experienced analog designer; that does not have a fixed circuit library but can be applied to a broad range of analog design problems; and that can synthesize variation-tolerant, high-performance circuits.

In this book, we present a new variation-tolerant analog synthesis strategy that we believe is a significant step towards ending the wait for a practical analog synthesis tool. We present a new synthesis strategy that can automate fully the path from a circuit topology and performance specifications to a sized variation-tolerant circuit schematic. This strategy relies on asymptotic waveform evaluation to predict circuit performance and simulated annealing to solve a novel non-linear infinite programming optimization formulation of the circuit synthesis problem via a sequence of smaller optimization problems. We have implemented this strategy in a pair of tools called ASTRX and OBLX. ASTRX is a circuit compiler that generates a performance prediction module that maps the component and voltage values in the circuit to the performance metrics specified by the user. More exactly, ASTRX generates code that implements a cost function that is carefully constructed so that its minimum value occurs at a circuit design that best meets the input specifications. This cost function code is then compiled and linked to OBLX, which uses simulated annealing to solve numerically for its minimum, thereby designing the circuit.

To show the generality of our new approach, we have used this system to re-synthesize essentially all the analog synthesis benchmarks published in the past decade;

ASTRX/OBLX has re-synthesized circuits in an afternoon that, for some prior approaches, had required months. To show the viability of the approach on difficult circuits, we have re-synthesized a novel, published, patented, high-performance operational amplifier; ASTRX/OBLX achieved performance comparable to the expert manual design. To test the limits of the approach on industrial-sized problems, we have synthesized the component cells of a pipelined A/D converter; ASTRX/OBLX successfully generated cells 2X-3X more complex than any published to date. Finally, to show the ability of our tools to create variation-tolerant circuits, we use them to design analog cells that are insensitive to operating range and manufacturing line variations; ASTRX/OBLX completed the novel synthesis task of designing a band-gap circuit that is stable over temperature and process variations.

This work is the culmination of more that five years of research in the Center for Electronic Design Automation (CEDA) at Carnegie Mellon University. The research groups within CEDA provided a forum to discuss the many ideas on which this book is based: Team Rob, the Analog Circuit Group, the ACACIA Group, the AWE Group, and in particular the optimization group that formed briefly for the summer of 1991 helped guide this research in a profitable direction. We would also like to thank the many people who have contributed to this work through direct guidance or simply via beneficial conversations over the years. Our thanks go to Bulent Basaran, Erik Carlson, Dennis Ciplickas, John Cohn, Peter Feldmann, Dave Garrod, Ramesh Harjani, Rajeev Jayaraman, John Lee, John Kibarian, Kannan Krishna, PC Maulik, Sujoy Mitra, Sudip Nag, Katsu Nakamura, Vivek Raghavan, Bob Stanisic, Professor Stephen W. Director, Professor Ignacio Grossman, Professor Ron A. Rohrer, and Scott Kirkpatrick.

ESO

CHAPTER 1 *Introduction*

1.1 Focus

This book describes a novel, *practical* approach to the *synthesis of high-performance analog circuits*. By an *analog circuit* we mean an analog cell, a circuit small enough that it would be designed as a single unit, typically an interconnection of a few tens of transistors, capacitors, and other fundamental circuit devices; by *synthesis* we mean fully automatic design to meet performance (rather than behavioral) specifications; and by *practical* we mean that the approach requires substantially less effort on the part of the designer than manual design yet yields comparable results.

To someone new to the area of analog synthesis, these definitions of analog synthesis and practical synthesis may seem somewhat surprising, since in digital design terminology, synthesis typically refers to translation from a *behavioral* description to a circuit implementation. However, in analog design terminology, synthesis typically refers to design of a circuit with a known qualitative behavior, such as amplification, to meet the performance requirements of the problem at hand. The reasons for this difference are described more fully later in this chapter. Similarly, although one might assume that any automatic design tool would be more practical than manual design, as we shall see, perhaps because of the inherent difficulty of the analog design problem, until recently researchers have had comparatively little success at creating an analog synthesis tool that is truly practical.

This chapter sets the context for the book by more completely defining the analog design problem and the part of it that practical synthesis of high-performance analog circuits addresses. It also provides a preview of some of the ideas discussed in greater detail in later chapters. In the balance of this chapter we motivate the need for automating the design of analog circuits, describe how cell synthesis fits into the analog design flow, present the goals of an analog cell synthesis tool, summarize our new approach to analog cell synthesis, preview a few synthesis results, and outline the remainder of this book.

1.2 Motivation

Many tasks that once were performed with analog circuitry are now performed by computers or at least with the aid of digital technology. However, despite the popular wisdom that computers can do anything, digital technology has not rendered—and will not render—analog circuitry obsolete. Liquid crystal displays, magnetic disk drives, and compact disc players are just a few of the modern marvels that most people consider part of the *digital* revolution that actually rely on a core of *analog* integrated circuitry. In fact, market pundits have predicted that the percentage of digital circuitry in an average electronic product has peaked [Collet 91]. Tomorrow's marvels— speech recognition systems, HDTV's, and ultra-small low-power personal communication devices—will contain an even greater percentage of analog circuitry.

Although new products will increasingly rely on analog circuitry to improve speed, decrease power consumption, or interface to the real (*i.e.*, continuous) world, industry is only partially prepared for this analog design renaissance. On one hand, integrated circuit (IC) fabrication techniques are poised to handle these new applications: entire mixed analog/digital systems that once required large, expensive circuit boards full of discrete components can now be implemented on a tiny inexpensive application specific IC (ASIC). On the other hand, the mismatch between the state of the art in *digital* computer aided design (CAD) tools and *analog* CAD tools hinders the resurgence of analog. With the advent of logic synthesis tools [Brayton 90] and semi-custom layout techniques [Kuh 90], digital CAD tools can quickly synthesize designs with literally millions of components, providing a rapid path to silicon for digital ASIC products. In contrast, most analog circuits are still designed manually by experts using time-consuming techniques that have remained largely unchanged in the past 20 years [Harjani 89a, Gray 87]. Further, although automating analog design has remained elusive, this full-custom analog design is often routine for accomplished human designers yet takes their valuable time away from more interesting and productive circuit- and system-level problems. Since reducing time-to-market is now crucial to making a profit, the lengthy design cycle due to this lack of analog CAD tools is an increasingly expensive liability in the design of mixed analog/digital ASICs. What is needed to mitigate this disparity between prevailing digital and analog design methodologies is an effective suite of analog CAD tools to bypass the analog design roadblock.

1.3 The Mixed-Signal Design Process

To better understand the difficulty of automating the analog design process, consider the simplified view of it shown in Figure 1.1. Here, the initial system-level concept is refined through a series of steps ending in fabrication. First, in *architectural design*, the system concept is decomposed into a collection of analog high-level circuit building blocks, that taken together will fulfill the system-level concept. These blocks are further decomposed until they are small enough to treat as atomic circuit cells. Then, in *cell design*, a detailed circuit-level schematic is created for each cell. In *cell layout*, the physical geometry within each cell is determined. Finally, in *system layout*, the global placement and routing that stitch the cells together is completed to re-create the architecture and fulfill the system concept.

Our focus is on cell design, since synthesis of high-performance analog cells is simply automatic cell design. We can further expand cell design into the sub-tasks shown in Figure 1.2. The inputs to cell design are the cell specifications as determined by the chosen architecture, the fabrication technology for which the cell is being designed, and the environment in which it will operate. The first task is to choose a topology, which is the appropriate interconnection of transistors, capacitors, and other circuit devices; the second is called *sizing and biasing* because circuit parameters such as device sizes and independent source biases are picked to meet the nominal specifications; and the final task is to modify these parameters to ensure the circuit is tolerant to operating environment and manufacturing line variations (*i.e.*, improve yield). The last two sub-tasks are the focus of this book, and we will discuss these tasks — and how they can be automated — in greater detail throughout the balance of this work.

Although the bulk of effort in CAD has focused on digital design, some of the earliest CAD research focused on the analog design problem. With the resurgence of demand for analog circuits has come a renewed interest in CAD tools intended to speed the analog portion of the mixed-signal ASIC design process shown in Figure 1.1 and Figure 1.2. However, until very recently, there has been only mixed success in automating the various components of that process. One success in this area is computer-based analog circuit simulation. As shown in the figures, simulation and verification have always been cornerstones of the analog design process because they are the gates between each of its steps. Computer-based cell-level simulation was the first component of the process in which significant success was achieved in analog CAD [Rohrer 92], and it is still the focus of much research. Further, the relationship of

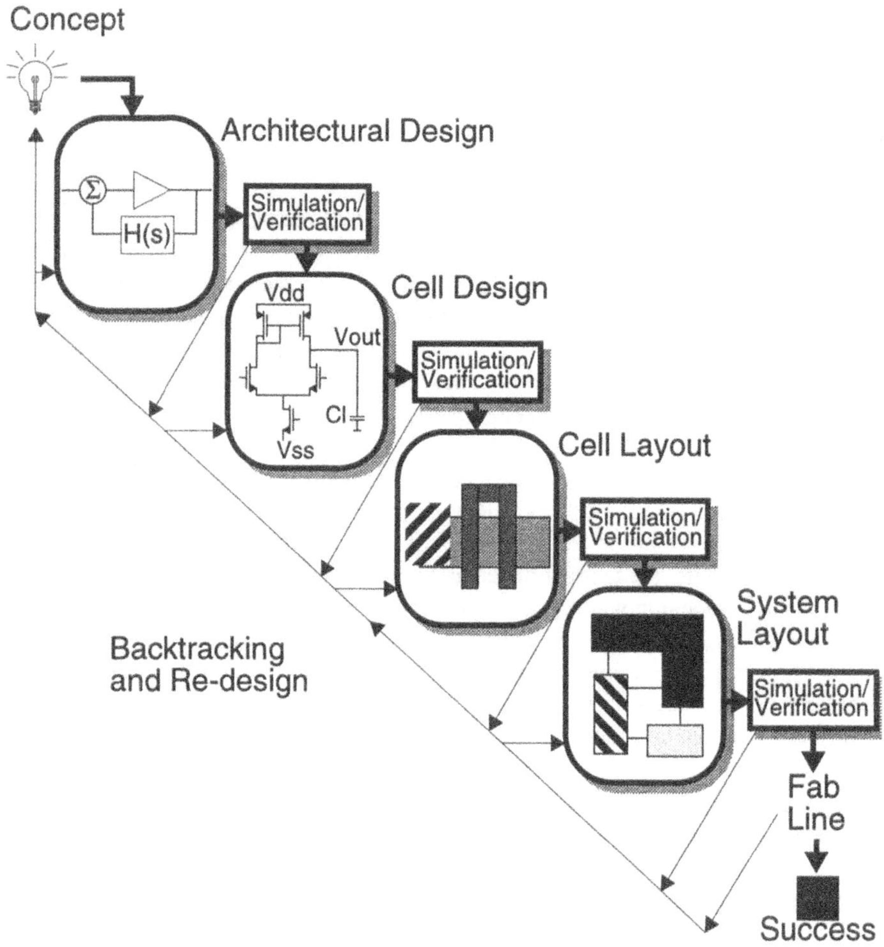

FIGURE 1.1 Simplified View of the Analog Design Process

simulation to synthesis is important: the same techniques used in simulation are used within synthesis tools themselves to predict and validate circuit performance during design. However, the research in this area is large and somewhat tangential to our own. There are several books and many areas of ongoing research in this area, for example [Stanisic 91, Getreu 89, Suyama 89, Saleh 90, Vlach 94, Raghavan 93, Rohrer 92].

In contrast, automation of the design steps in Figure 1.1 and Figure 1.2 has been less successful. Although researchers have created a variety of tools, most have been cumbersome to use and have not dealt adequately with the analog nature of mixed-signal design, making them unacceptable alternatives to manual design in performance-oriented industrial environments. However, with the advent of the cell design tools described in this book, and analog-sensitive automatic cell layout [Garrod 88, Cohn

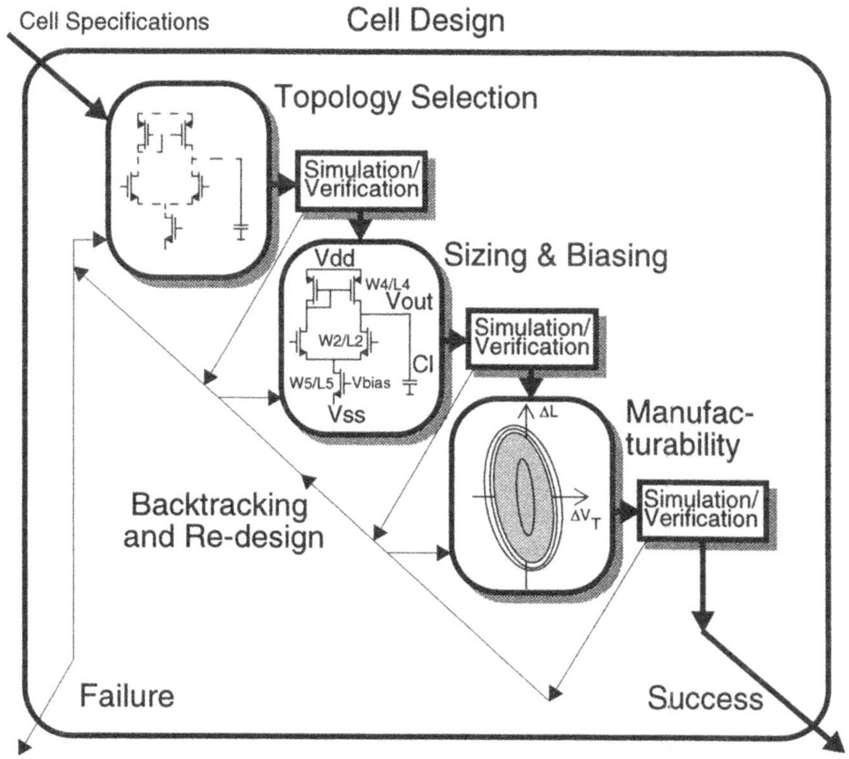

FIGURE 1.2 Sub-tasks in the Analog Cell Design Step

91, Basaran 93], system-layout [Mitra 92], and power distribution [Stanisic 93], the prospects for more fully automating these tasks are improving.

Analog CAD has been less successful than digital CAD for a number of reasons. One of these reasons becomes apparent when the process a real analog designer goes through is compared to the simplified model in Figure 1.1 and Figure 1.2. Despite what the regularity of the figures may suggest, the analog design flow is rarely a confident progression through discrete, formally verified design steps to successful first fabrication. A real analog designer does not blindly perform each of the various steps in isolation without regard for what task comes next. In analog design, even in the most abstract phases of design, such as architectural design, the designer must often consider cell-design and layout-level concerns [Hauser 86, Castello 85]. Moreover, to a far greater degree than in digital design, skilled analog designers implicitly consider the effect decisions at the current step in the design will have in subsequent steps. Because of the large number of interrelated design concerns and the difficulty of isolating the effect of a decision to a single design step, even the best designers must often backtrack and reconsider failed decisions. This holistic approach to design is required within the cell design step, Figure 1.2, and although the designer focuses in turn on topology selection, sizing and biasing, and yield improvement, it is more accurate to say these three aspects of cell design evolve concurrently.

In contrast to this blurring of steps in the analog design process, the digital design hierarchy and its consequent information hiding are universal principles that greatly simplify digital design. This is possible because the designer can make assumptions about successive design steps. For example, for system-level digital design it is usually sufficient to model complex logic functions with a boolean equation and a simple delay and/or power parameter, ignoring the detailed design of the logic circuits until later in the design process. This use of hierarchy provides a great deal of leverage because system-level digital designers need not consider or even understand the transistor and layout level design issues. In digital design, if the process changes, it is usually sufficient to simply update the delay and power consumption parameters. Although hierarchy is still very useful in the analog design process, learning which details can and cannot be abstracted is one of the most difficult analog design skills to master and is part of the lore that makes each skilled designer unique. Consequently, hierarchy is not the strong, generally applicable set of principles in analog design that it is in digital design. This inability to apply a universal set of abstractions considerably complicates the task of constructing general-purpose analog synthesis tools.

A second unfortunate implication of this inability to hide information is that the analog designer must consider many more performance specifications than her digital coworker, making the use of semicustom design techniques such as libraries of standard cells far less powerful in the analog domain. Since analog cells typically have in excess of 20 performance specifications, frequent process shrinks make it impractical to maintain the several million cells required to cover a useful cross-section of specifications [Bassak 95]. The most desirable solution is to synthesize analog circuits on demand to exactly match the required performance specifications.

1.4 Goals of Analog Design Automation

Broadly speaking, automating the entire analog design process is the goal of CAD research in this area [Rutenbar 93], and synthesizing analog cells on demand to exactly match the required performance specifications is the goal of the work described in this book. In this section we describe a more detailed set of goals that we then use to create evaluation criteria to measure the practicality of a given approach to analog design automation. Many of these goals apply to any of the tasks in Figure 1.1, but in this book we will apply them primarily to analog cell synthesis. These goals include:

1. *Shortened design times and reduced design costs:* Manual design is a long and often tedious task. Automation will considerably improve productivity and shorten time-to-market. Moreover, improved productivity will reduce the amount of non-recoverable engineering effort, often the largest component of cost for a design. This is particularly important when a mixed-signal ASIC is competing with an existing board level design for which the analog functionality, while more expensive to produce, has already been designed.
2. *Improved design quality:* Designing analog systems and cells is a large but detail intensive task, error-prone for human designers, and an ideal target for automation. A practical synthesis tool must create designs that are more likely to perform correctly the first time than those designed manually. Also, there should be little or no penalty for using synthesis: parametric performance of the automatically designed cell should match the performance attainable by full-custom design.
3. *Decreased production costs through increased manufacturability:* For the synthesized design to be manufacturable, the automatically designed circuit should be insensitive to variations around the nominal point in both the fabrication line and the circuit operating environment. By synthesizing for manufacturability as well as nominal performance, synthesis tools can enhance the manufacturing yield of circuits, thereby improving profitability.

4. *Independence from fabrication processes:* In this day of tight silicon foundry space and ever-shrinking minimum device dimensions, the ASIC designer cannot afford to target her design for one particular fabrication line. If the synthesis tool does not make process-dependent assumptions and takes the process description as input, mapping previously synthesized designs on to the latest fabrication process is as straight forward as re-running the tool.
5. *Simplified design process:* If many of the detailed concerns can be taken care of automatically, the design engineer can focus on larger system level issues instead of being consumed by minutiae. As a consequence, designers at all levels of proficiency will be able to design standard analog circuits. However, an automatic system that is difficult to learn and use may reduce time-to-market, but it will never be popular among the engineers and will be unlikely to be adopted as an integral part of the analog design process.

The goals detailed above drive the work described in this book. To provide more quantifiable measures of success at achieving these goals, we define four metrics that codify the most important criteria in the above goals but are more concrete and related to the measurable characteristics or behavior of an analog synthesis tool. These metrics are automation, accuracy, breadth, and variation tolerance, which we define as follows:

- *Automation*: the ratio of the time it takes to design a circuit for the first time manually to the time it takes with the synthesis tool. This is perhaps the most important metric because it captures the goal of shortened design times and reduced design costs. When comparing synthesis tools, manual design time will be the same for a given circuit, so maximizing automation is equivalent to minimizing the sum of *preparatory effort* and *synthesis time*. Preparatory effort is the designer-time/effort required to render the circuit problem in a form suitable for input to the tool and synthesis time is the CPU time required to complete the circuit design. For example, for synthesis tools that execute a circuit-specific plan to complete a design, if the required circuit is unlike those the system has designed previously, the preparatory effort alone will take *several times longer* than completing the single design manually, so these kinds of tools do a poor job automating new circuit design. Such tools may still be useful, but only if there are a large number of very similar analog design tasks. Thus, synthesis time alone is not an adequate indicator of automation.

- *Accuracy:* the discrepancy between the synthesis tool's internal performance prediction mechanisms and the actual performance of the completed circuit. Since layout effects are not considered in the remainder of this book, the performance predictions of a detailed circuit simulator that uses realistic device models are accepted as the actual performance of the circuit.

  Accuracy reflects the goal of improved design quality. Assuming synthesis tools always find the circuit solution with the best performance given their ability to model that performance, accuracy determines the overall design quality. Of course, the ability to find the optimal solution will depend on the search method used, but optimality is almost impossible to prove for a task as complex as analog design. Consequently, we rely on accuracy as our most objective indicator of design quality.

- *Breadth:* the set of analog design problems that the tool can complete. Breadth is much more difficult to quantify than automation or accuracy but is important because it indicates how likely a tool is to be applicable to the design task at hand. We define a system's breadth by using three dimensions: *generality*, *complexity*, and *scope*.

  Generality is determined by the classes of circuits and types of performance specifications that can be successfully handled by the synthesis tool. For example, a tool may not be able to handle non-linear specifications or may design only operational amplifiers.

  Complexity is determined by the largest circuit synthesis task that can be successfully completed by the synthesis tool. This is somewhat difficult to quantify because the complexity of a synthesis task can be affected by many factors including the number of designable parameters (element values and device sizes), the number and difficulty of the performance specifications, the number of components in the circuit, and the inherent difficulty of evaluating the performance of the circuit.

  Scope is determined by the fraction of the analog design flow that is addressed by the tool. Of course, we are interested primarily in the cell design step, which is composed of the sub-tasks in Figure 1.2. At present no synthesis system of which we are aware addresses all three tasks, and the work described in this book is the only system that combines sizing and biasing and manufacturability into a single formulation.

- *Variation Tolerance:* The ability of the tool to create analog circuits that are tolerant of manufacturing line and operating point variations. This metric addresses both the goal of improved design quality and the goal of decreased production costs through increased manufacturability. We quantify variation tolerance as the combination of *manufacturing yield* and *worst-case performance over an operating range*. The manufacturing yield is the ratio of the number of circuits that meet behavioral and parametric performance specifications to the total number of manufactured circuits, despite fluctuations in temperature, mask alignment, and other properties of the manufacturing line. Worst-case performance over an operating range measures the circuit's insensitivity to variations in its operating environment, such as the supply voltage and ambient temperature. For specifications where a parameter must be above or below a certain threshold, both the worst-case and nominal values must be on the correct side of the threshold for the specification to be met, and, for specifications where a parameter is being maximized or minimized, the worst case value over the operating range is used as the achieved performance.

An ideal system maximizes each of the metrics described above: automation, accuracy, breadth, and variation tolerance. The work presented in this book focuses on a tool that automates analog cell design and attempts to reach the above goals by maximizing the four metrics.

1.5 Research Direction

The work presented in this book is a CAD tool for the practical synthesis of high-performance analog circuits, which means it addresses the goals outlined in Section 1.4. Tools of this type, known as *analog circuit synthesis systems*, promise automatically-designed analog cells that match full-custom performance and are manufacturable, yet require substantially less than full-custom design time. Unfortunately, early predictions of quick success with fully automated analog design proved premature [Rohrer 92]. As we discussed in Section 1.3, the inherent difficulty of analog design is substantially to blame for this lack of success. Although researchers have proposed several frameworks for such tools, few have migrated from research institutions to industrial design houses, and none have proven practical and effective enough to be accepted as an integral part of the industrial mixed-signal ASIC design process.

Our goal is to address directly the problems that have prevented this transition from research to practice. To do so, we must overcome two challenges. The first of these is the prohibitive one-time effort required to derive the complex equations that drive other synthesis tools, and the second is to create a unified synthesis formulation that can design variation-tolerant circuits. For ease of presentation, in the remainder of this chapter and throughout the balance of the book, we consider the nominal synthesis problem first, then build a unified synthesis formulation upon that foundation.

1.6 A New Nominal Synthesis Approach

Our first challenge is to address the automation and accuracy problems that arise in "equation-based" synthesis tools. In this section, we introduce a new nominal synthesis approach that addresses these problems. This is a nominal technique because manufacturability issues are not considered, these will be taken into consideration in the variation-tolerant unified synthesis formulation that addresses our second challenge. The ideas and formulation introduced in this section are described completely in Chapter 3, Chapter 4, and Chapter 5.

Because they rely on a core of equations, we refer to all previous approaches to synthesis as "equation-based", and discuss their architecture in terms of the simplified search loop shown in Figure 1.3. With each pass through this loop, a search mechanism perturbs the element values, transistor dimensions and other variable aspects of the circuit in an attempt to arrive at a design that more closely meets performance specifications and other objectives. Performance equations are used to evaluate the new circuit design, determine how well it meets its specifications, and provide feedback to the search mechanism to guide its progress. Even for a relatively simple analog circuit, these performance equations are very complex, require considerable analog design expertise to derive, and must be entered as thousands of lines of program code. This has a significant negative impact on automation, since for a textbook design this process can take weeks [Koh 90], and for an industrial design it can take several designer-years [Degrauwe 89]. The arduous and time-consuming task of carefully working out and testing performance equations must be duplicated each

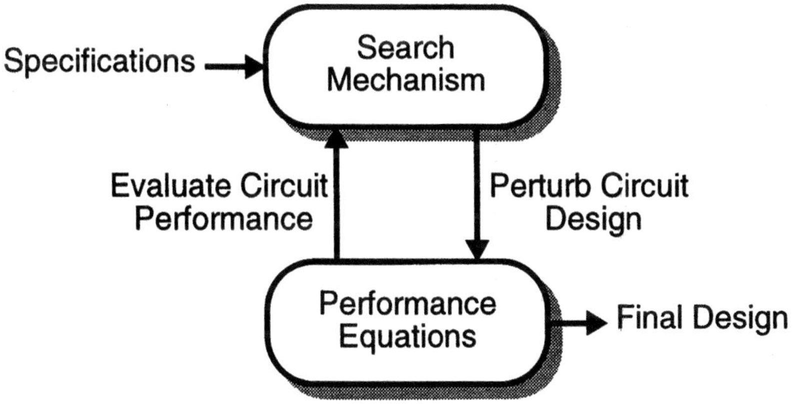

FIGURE 1.3 Search Process Used in Equation-Based Analog Synthesis Tools

time performance equations for a new circuit topology are added to the synthesis tool's library. As a result, it is almost always more efficient for an industrial mixed-signal ASIC designer to design circuits manually rather than dedicate the effort required to teach these tools to design the circuit "automatically".

This does not mean that researchers are unaware of this barrier to the acceptance of their analog synthesis tools. There have been several variations on the simple template of Figure 1.3 in an attempt to make it easier to codify the performance equations. Our experience with the synthesis tool OASYS [Harjani 89] parallels the evolutionary path followed by many other synthesis tools [Degrauwe 89, Harvey 92] that have evolved strategies to confront this problem of acquiring circuit design expertise, although in our case we eventually concluded that the best solution was to re-design completely the equation-based analog synthesis methodology. As in most early synthesis tools, in the original implementations of OASYS [Harjani 87] the distinction between the circuit-specific knowledge in the system and the circuit-independent architectural aspects of the system was only intellectual. The first step towards allowing new circuits to be added was to implement a circuit independent architecture that would remain unchanged as different circuits were added. We accomplished this with OASYS VM [Ochotta 89a]. The next step in our evolutionary plan was to allow the circuit-specific knowledge to be expressed in a form that was simpler and more familiar to analog circuit designers than a general-purpose programming language. This was the design goal of the OASYS analog design language (OADL) [Ochotta 89]. However, during the language design process and through extensive feedback from circuit designers, it became increasingly clear that fundamental shortcomings of the equation-based synthesis methodology would eventually doom the equation-based approach outside the research community. The most crucial of these shortcomings manifest themselves as problems in accuracy and automation. We summarize these problems as follows:

- **Accuracy:** Equation-based approaches rely heavily on simplifications to circuit equations and device models. The need for high performance circuits using technologies with submicron devices invalidates the use of many of these simplifications. For example, in a 3μ MOS process, $I_{DS} = K'W/2L(V_{GS} - V_T)^2$ is a workable model of the current-voltage relationship for a device and can be inverted to allow either voltage or current as the independent variable. But this is a grossly inaccurate model for a device with a submicron channel length. The need to support complex device models and circuit design goals that push the limits of these models is fundamentally at odds with equation-based strategies that rely on these simplifications.

- **Automation:** Equation-based synthesis tools appear to design circuits quickly. But, the stated run-time of the tools is very misleading because it does not consider the time required to derive the circuit equations. Including this derivation time, equation-based synthesis tools require substantially more effort for a first-time circuit design than the already time-consuming manual design process. With time-to-market the overriding concern, it is unlikely that designers can afford the time to teach equation-based synthesis tools to do routine cell design.

Other researchers have pursued methods similar to OADL in an attempt to minimize this "teaching time", and a more complete discussion of these methods appears in Chapter 2. Despite these efforts, we are unaware of any previous approach to analog synthesis that has effectively addressed these accuracy and automation concerns.

One beguilingly simple solution is to replace the equations with a detailed simulation technique. This is the basic approach that was first proposed for analog circuit *optimization* decades ago [Rohrer 67], rediscovered when faster computers and improved simulators made it practical for research [Shyu 88, Nye 88, Hocevar 90], and is now making its way into industrial CAD systems. Because detailed simulators are used, no designer-supplied equations are required and the performance prediction is very accurate. Unfortunately, because of the simulation time, these simulation-based optimizers are painfully slow when applied to synthesis. Although computation speeds continue to increase, what is needed is an algorithmic innovation that can yield the accuracy and generality of conventional simulation in a fraction of the execution time.

By combining five key ideas, we have designed an alternative approach to analog synthesis that directly addresses the accuracy and automation problems of previous approaches. Based on the goals for analog synthesis identified in Section 1.4 and a critical analysis of existing analog design automation tools, we have set three specific targets for our nominal synthesis tool. First, it should take only hours of preparation to design a new circuit and not require weeks or months of circuit analysis. Second, it should yield accurate performance predictions for high-performance circuits and not suffer from errors and simplifications in device models or circuit performance equations. And third, it should find high-quality circuit design solutions without regard to starting point and not require the designer to provide a good initial circuit design as a starting point for an optimization process or easily become trapped in a poor local minima.

Our primary goal here is efficiency. We wish to streamline the path from a circuit idea to a sized circuit schematic. In gauging efficiency, it is important to distinguish the run-time of the tool from the overall time required to design a new circuit. Equation-based synthesis tools run very quickly but require the designer to perform the arduous and time-consuming task of deriving, coding, and testing performance equations. We believe a designer's time is more valuable than a computer's time and are willing to exchange several more hours of computer effort for several months of designer effort.

To achieve these goals our nominal circuit synthesis strategy relies on five key ideas. These ideas and their implementation are described completely in Chapter 3, Chapter 4, and Chapter 5, but can be summarized as follows:

1. A non-linear constrained *optimization formulation* of the circuit synthesis problem that we map to an unconstrained optimization algorithm and solve numerically.
2. A *relaxed D.C. formulation*, that avoids a CPU intensive D.C. operating point solution after each step in the optimization.
3. *Simulated annealing* [Kirkpatrick 83] as the optimization engine to provide robustness and the potential for global optimization in the face of many local minima.
4. *Asymptotic Waveform Evaluation* (AWE) [Raghavan 93, Pillage 90] augmented with some simple, automatically generated analytical equations to predict circuit performance.
5. A compiled database of industrial quality, non-linear device models, called *encapsulated device evaluators*, that provides the accuracy of a general-purpose simulator while making the synthesis tool completely independent of low-level device modelling concerns.

We combine these five ideas to create an architecture that provides a fully automated path from an unsized analog circuit topology and a set of performance specifications to a completed, synthesized circuit (see Figure 1.4). This path is comprised of two phases:

- **Compilation:** Compilation generates a performance prediction module (replacing the performance equations of Figure 1.3) that maps the component and voltage values in the circuit to the performance metrics specified by the user. More exactly, code is generated to implement a cost function whose minimum value occurs at a circuit design that best meets the input specifications. Because of our relaxed D.C. formulation, the D.C. correctness constraints that will enforce Kirchhoff's laws for the input topology are automatically derived and encoded in this cost function. Finally, to evaluate this cost function, the appropriate links to the encapsulated device evaluators and AWE are also compiled in.

- **Solution:** This cost function code is then compiled and linked to our solution library, which uses simulated annealing to solve numerically for its minimum, thereby designing the circuit.

These five key ideas and the resulting two-phase architecture have been called "equation-free" synthesis, to emphasize the clean break with equation-based synthesis approaches [Ochotta 92].

As presented so far, this nominal approach has some important limitations that we now enumerate. The approach is intended to automate primarily the nominal sizing and biasing of analog circuits whose predominate operating characteristics are linear or small-signal. The approach as we initially describe it has important limitations when designing circuits with important large-signal performance specifications, designing or selecting circuit topologies, and designing circuits that are tolerant of variations in operating conditions and manufacturing. However, we can overcome these limitations as we describe below.

The first limitation concerns large-signal and transient performance specifications. Large-signal and transient performance must, in general, be evaluated with either

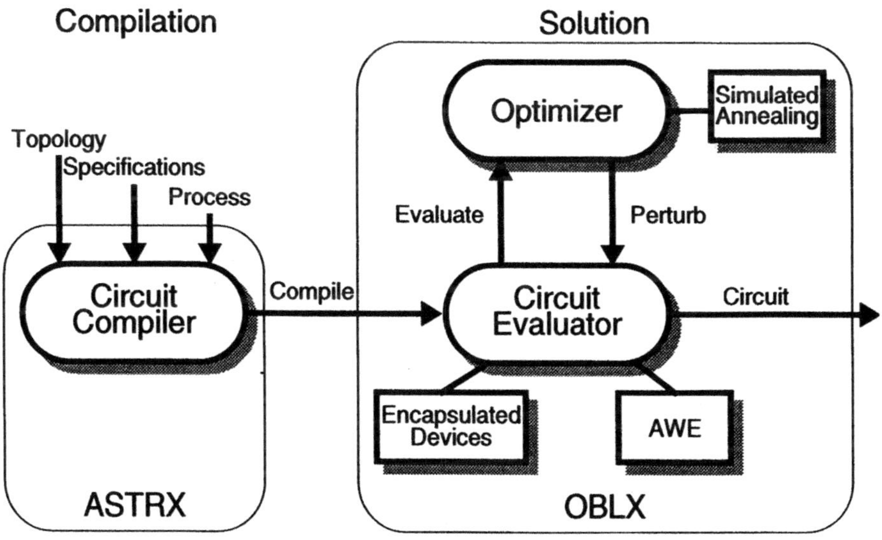

FIGURE 1.4 New Synthesis Architecture

simulation or designer-supplied equations. Because our approach relies on AWE, it can determine linear or small-signal performance over frequency extremely efficiently, but cannot directly handle transient or large-signal specifications. Other existing simulation techniques that can handle other kinds of performance characteristics are too slow for synthesis. To allow these other performance characteristics to be modelled, we support the addition of designer-supplied equations to the automatic analyses provided by AWE. Fortunately, most important large-signal specifications are adequately modelled with simple equations that can readily be written by inspection of the circuit. Thus, AWE replaces the most burdensome equations while simpler ones are left to the designer. Although this is contrary to the "equation-free" philosophy, in practice it is not a large impediment to automation. As AWE and other simulation techniques evolve, it may be possible to replace even large-signal equations with very rapid transient simulation [Alaybeyi 90, Visweswariah 91].

A second limitation in the approach as formulated thus far is that it sizes and biases a single fixed circuit topology. Although some existing equation-based synthesis tools presently include support for topological design, we choose to approach first the more fundamental sizing and biasing problem. We feel that an effective sizing and biasing methodology must form the core of any complete synthesis system. Topological design is almost always the process of selecting from a small set of known topological alternatives, and even if this selection is done manually it can be greatly sped up with automatic sizing and biasing. Sizing and biasing are the most tedious tasks in the analog cell design process, making them the ideal first target for automation. When our new core technology is proven, it can then be augmented with methods for automatic topological design. Many existing methods for topology selection [Koh 90] can easily be used as a pre-processing step to select a suitable topology that will then be sized using the techniques described here. Alternatively, our nominal approach does not preclude simultaneous topology selection and sizing (*e.g.*, the method recently developed in [Maulik 92]), and preliminary work on extending the approach in this direction is encouraging [Jones 96].

The third limitation is that the nominal approach does not produce circuits that are tolerant of variations in the manufacturing process or in their operating environment. To be truly practical, a synthesis tool must create circuit designs that can be used, so we must extend the approach to consider these variations. This is the second challenge addressed by the work in this book, and we defer further discussion until Section 1.8.

1.7 Preview of Nominal Synthesis Results

To show the viability of our new nominal approach, we have implemented its two phases in a pair of tools called ASTRX and OBLX. Given little more than a simulation deck, these tools can accurately size and bias a new circuit topology to meet user-supplied performance specifications. Prior approaches required up to 3 orders of magnitude more time to enter a new circuit description, yet predicted performance less accurately when compared with detailed circuit simulation results. For example, it took a design team 9 man-months to encode an OTA-style operational amplifier into the equation-based tool IDAC [Degrauwe 89], yet it required only an afternoon to complete the entire design with ASTRX/OBLX. For the recent equation-based tool STAIC [Harvey 92], it required about 10,000 lines of code in its specialized language to describe three circuits, yet the circuit shown in Figure 1.5, which is larger than their most complex example, requires only 119 lines to describe in ASTRX/OBLX—the majority of which describe the circuit topology. Finally, equations for OPASYN [Koh 90] took only about 2 weeks per circuit to derive, but differ from simulation results by as much as 200%. Predictions from ASTRX/OBLX using AWE are typically identical to SPICE-like detailed simulation.

However, single case comparisons do not tell the entire story. It is perhaps more important that ASTRX/OBLX performs consistently well on many different circuits, and we verify this with a suite of results in Chapter 6. The limited range and performance of the circuit topologies published for prior systems is perhaps the best indica-

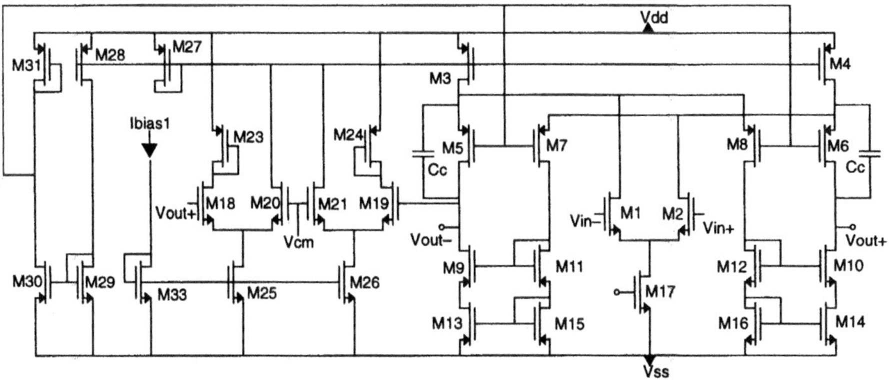

FIGURE 1.5 Novel Folded Cascode Circuit Automatically Designed with ASTRX/OBLX

tor that the first-time effort to design a circuit has always been a substantial barrier to obtaining a broader range of results. Because of the effectiveness of ASTRX/OBLX in automating the design process, we present results for a suite of analog circuits that vary widely in performance and complexity. For example, Figure 1.5 is a unique high-performance fully differential folded cascode operational amplifier designed by ASTRX/OBLX. This circuit is a unique case for several reasons. First, it is not a text-book design and the performance equations cannot simply be looked up. Second, the performance of the circuit is very difficult to express analytically and as many as 6 poles and zeros may non-trivially affect the frequency response near the unity gain point. Finally, the circuit is a novel high-performance design published in [Nakamura 92] and as such is a significant test for any synthesis tool. Further, we have access to the device sizes originally obtained by an expert designer and the original process parameters for which it was designed. Table 1.1 is a comparison of a re-design of this circuit using ASTRX/OBLX with the highly optimized manual design as originally published for the same 2μ process. Surprisingly, ASTRX/OBLX actually finds a

TABLE 1.1 Comparison with Manual Design for Circuit Novel Folded Cascode

		Automatic Re-Synthesis with ASTRX/OBLX		
Attribute	**Manual Design**	**Specification**	**Design**	**Simulation**
Cload	1pF	1pF	1pF	1pF
Vdd	5V	5V	5V	5V
D.C. gain (dB)	71.2	≥ 71.2	82	82
gain bandwidth (MHz)	47.8	maximize	89.3	89.0
phase margin (°)	77.4	≥ 60	91	91
PSRR (Vss)	92.6	≥ 93	112	112
PSRR (Vdd)	72.3	≥ 73	77.3	77.2
output swing (V)	± 1.4	± 1.4	± 1.4	1.3
slew rate (V/μs)	76.8	≥ 76	92.5	87
active area ($10^3\mu^2$)	68.7	minimize	56.5	
static power (mW)	9.0	≤ 25.0	12.1	12.1
time/ckt. eval (ms)			83	
CPU time (hrs. / 5 runs)			9.7	

design with higher nominal bandwidth at the cost of less area. Although we are pleased with the ability of OBLX to find this corner of the design space, this does not mean that ASTRX/OBLX out-performed the manual designer in all respects. In fact, the expert designer was willing to trade nominal performance for better overall yield and performance over varying operating conditions. This result shows the ability of ASTRX/OBLX to handle difficult nominal circuit design problems, but also highlights the need for variation-tolerant synthesis.

1.8 A New Unified Synthesis Formulation

Results from the new nominal approach to analog synthesis as implemented in ASTRX/OBLX, are better in terms of automation, accuracy, and breadth than any other existing synthesis tool. However, as the nominal synthesis result for the circuit in Figure 1.5 shows, the nominal approach by itself is not a practical solution because the circuits it designs are not tolerant of variations in the manufacturing process and in the circuit operating environment. To address variation tolerance is the second challenge that must be met in order to create a practical high-performance analog circuit synthesis system. In this book we present a unified formulation for analog synthesis that does address these concerns and overcomes the practical limitations of prior nominal synthesis approaches. In addition to the general goals for circuit synthesis that were outlined in Section 1.4, we have four specific goals for extending the nominal architecture to be variation-tolerant. These are:

1. Address the fact that in the real world a practical circuit design meets specifications over a range of operating points, not just a single operating point.
2. Account for the inevitable fluctuations in a fabrication line, both from one die to another and from one device to another within a single die.
3. Use only one set of variables in both the nominal design and the statistical IC design.
4. Treat the above goals in an unified numerical formulation.

To achieve these goals, our variation-tolerant circuit synthesis strategy adds the following three key ideas to the nominal approach of Section 1.6:

1. Introduction of *range variables* to model supra-nominal variations. These variations include the circuit environment operating point fluctuations and the variations in the fabrication line.

2. Formulation as a *Non-linear Infinite Programming* problem to completely model the circuit design problem as a circuit synthesis problem.
3. Solution of this formulation with a sequence of two alternating optimization problems: a *nominal synthesis* problem solved using the techniques described in Chapter 3, Chapter 4, and Chapter 5; and a *worst-case optimization* problem solved using annealing.

This extended architecture, with the addition of these three key ideas, efficiently handles operating range and manufacturing line variations, creating a practical synthesis system for high-performance analog circuits. The key to the approach is to *completely* model the design problem as a mathematical program. This is because optimization-based approaches work only when they have the complete model of the design problem they are attempting to solve. The critical point of failure in previous attempts at synthesis was incomplete modeling of the circuit design problem as a synthesis problem. Such an incomplete and inaccurate model leads to synthesis results that do not meet any of the goals we specified for synthesis in Section 1.5.

The first step in completely modeling the circuit design problem is to introduce variables that can model the issues that are important to that design problem. In our case, the issues that need to be modeled are variations in the circuit environment and in the fabrication line used to manufacture the circuit. In this work we do this using a range variable. This is a distinctly different kind of variable than the one we have intuitive knowledge: circuit design variable. We will clarify this difference in Chapter 9.

The *non-linear* nature of integrated circuit devices such as transistors leads to a non-linear optimization-based formulation, as in the case of previous approaches. In Chapter 2, we formulate the nominal synthesis problem as a non-linear programming problem. Unlike this approach, the manufacturable synthesis approach requires performances to be met over a range of conditions, with an infinite number of points in a continuous range, so it must be formulated as a non-linear *infinite* programming problem.

The complete non-linear infinite programming model of the design problem is unfortunately not addressed by standard off-the-shelf optimization packages. We use a *sequence of alternating optimization sub-problems* in the style of [Eaves 71]: the state-of-the-art optimization-based nominal synthesis approach described in the first part of this book is used to do the actual synthesis, with a second optimization providing the worst case points which the nominal synthesis should consider. More exactly, the synthesis algorithm is synthesizing one circuit (*i.e.*, one set of device sizes and biases) to meet the specified performance at several different worst case points. One of these points is the standard nominal synthesis point. In addition, other worst case points are added by the worst case point optimization process "on the fly" during the synthesis process. A worst case point is a point in the space of manufacturing variables. This is in contrast to a design point, which is a point in the space of design variables. At a worst-case point, the current design point fails to meet the desired performance by the largest margin, hence the title "worst-case". The addition of worst case points to an interactive design approach implies that the performance evaluation evaluates the same design point at multiple worst case points to determine the circuit performance. As the dual-optimization approach evolves, more and more worst case points are identified, and considered by the optimization-based synthesis. A new worst case point is just like a new constraint. Prior to the addition of the first worst case point there was a single constraint: for instance, gain at the nominal point needs to meet its specification; after adding the new worst case point, we have two constraints: the gain at the nominal point *and* the gain at the worst case point, and *both* need to meet the gain specification.

Worst case points are added only when the design specifications at the current synthesis point are not met. This provides two valuable features: First, the performance evaluation is done at a minimal number of points, reducing the total time taken by the synthesis system. Second, this approach ensures that the final circuit indeed meets the design specifications across the entire range of circuit operation and manufacturing variations.

1.9 Preview of Variation-Tolerant Synthesis Results

Now that we have an understanding of the goals and strategy of the variation-tolerant synthesis approach, let us take a quick look at an example that highlights the ability of our strategy to meet its goal. The result we present here is that of a band-gap reference circuit used commonly to generate temperature-insensitive bias currents for many analog designs, including the operational amplifier. Unlike the typical amplifiers that have been synthesized by almost all prior approaches to synthesis, this is the first synthesis result for a band-gap reference circuit.

This design was generated in about a half-hour from a description that is little more than a circuit simulation deck for the schematic shown in Figure 1.6, highlighting the fact that the new formulation still provides excellent automation. Analysis of this circuit shows that ideally the current I_{out} behaves as $T^{0.5}$. As shown in Figure 1.7, simulation of the resulting synthesized circuit shows that the output current is able to match this behavior (the temperature exponent obtained from regression analysis is 0.42). In addition, post-synthesis full circuit simulation and the simulation used within synthesis match to about 7%. Therefore our approach can attain our goals of accuracy and full custom design quality. Our use of bipolar and MOS transistors

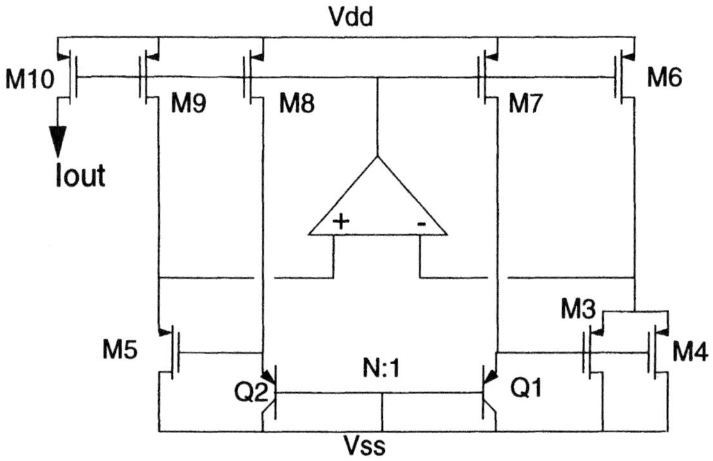

FIGURE 1.6 Simple Band-gap Reference Circuit

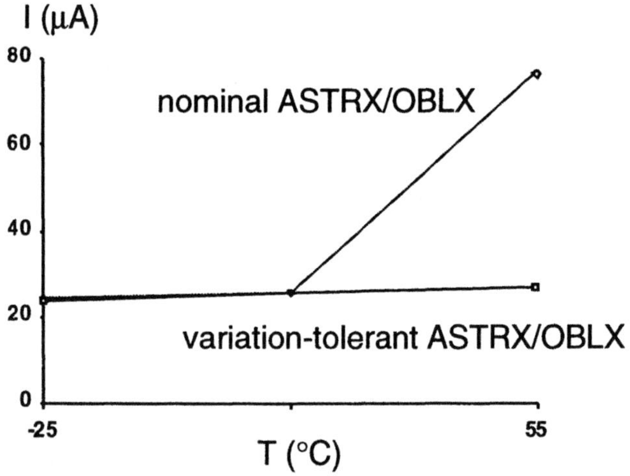

FIGURE 1.7 Temperature Sensitivity of Band-gap Reference in Figure 1.6.

within the same design shows technology independence. Finally, our ability to handle temperature ranges shows that we have a viable approach to analog cell synthesis beyond just a single nominal operating point.

1.10 Book Organization

The balance of this book presents in detail our practical approach to the synthesis of high-performance analog cells. The discussion is presented in two parts. First, the nominal synthesis strategy is presented in detail. Second, building on the nominal synthesis foundation, the variation-tolerant strategy is presented. We have organized the discussion as follows:

- In Chapter 2, we examine more extensively the previous approaches to the nominal analog synthesis problem. We begin with an overview of recent attempts to automate analog circuit design at the cell level. The balance of the chapter focuses on equation-based synthesis techniques, including a comparison of characteristics of several recent analog synthesis systems and the attempts to address accuracy and automation problems.
- Chapter 3 discusses our nominal circuit synthesis approach in detail. Its five key components—synthesis via unconstrained optimization, solution of the D.C. node voltages through optimization, simulated annealing, asymptotic waveform evaluation, and encapsulated device evaluators—are presented and their interrelationships are explored. The chapter concludes with a description of the two-phase architecture—compilation followed by solution—that is the key to combining these components into an effective synthesis tool.
- Chapter 4 explores the solution phase of this architecture. This phase implements the automatic design of the circuit by minimizing a carefully derived multi-dimensional cost function. The minimization algorithm, simulated annealing, is presented in detail and each of the components of our constrained optimization formulation is discussed.
- Chapter 5 explores the compilation phase of the two phase-architecture introduced in Chapter 3. Here, the problem description as presented by the user is translated into the multi-dimensional cost function used during the optimization phase. Although compilation precedes solution in the synthesis process, for clarity we discuss the compiler only after we have detailed the contents of the cost function it generates.

- Chapter 6 is the first of two chapters that present nominal synthesis results from our implementation of our nominal synthesis approach in the tools ASTRX and OBLX. Here, the focus is on comparing ASTRX/OBLX with prior approaches based on accuracy, automation, and breadth. In this process we will explore the breadth of circuits and performance metrics that can be designed for by ASTRX/OBLX.
- Chapter 7 presents a suite of additional circuit synthesis results generated automatically by ASTRX and OBLX. In this chapter we focus on validating the engineering and design decisions described in earlier chapters.
- In Chapter 8, we leave the realm of nominal synthesis and enter the realm of manufacturing process and operating environment variations. We begin the chapter with a close look at the weaknesses that render nominal synthesis impractical. We then review previous work on parametric yield optimization, a field closely related to variation-tolerant synthesis.
- Chapter 9 presents our new non-linear infinite programming formulation of the circuit synthesis problem. The chapter begins with the introduction of range variables and continues to present the formulation. A circuit design example is used to further clarify this formulation.
- Chapter 10 discusses the solution of the non-linear infinite program. After a short review of non-linear programming literature we begin a conceptual explanation of the solution approach we use. This is further clarified using a simple circuit example that shows how a sequence of sub-problems can be used to solve the complex overall problem. The balance of the chapter includes discussion of the interaction between these sub-problems, implementation details, and a synthetic example to show that our implementation works as described in the conceptual example.
- Chapter 11 presents the variation-tolerant analog synthesis results. To highlight the critical need for manufacturable synthesis, we compare each synthesized result with its corresponding nominal synthesis result.
- Finally, Chapter 12 concludes the book by enumerating the contributions of this work and suggesting directions for further research.

CHAPTER 2

Methods for Nominal Analog Circuit Synthesis

To understand fully the rationale behind our nominal approach to the synthesis of high-performance analog cells, it is important to discuss the existing attempts to automate analog design and the shortcomings of these attempts. In Section 1.3 we presented a simplified view of the mixed analog/digital design process, and highlighted the cell design step as the focus of our design automation work. In this chapter we discuss four general techniques that have been used to automate the cell design step: *layout-based design automation*, which attempts to build a circuit by snapping together pre-designed cells or portions of cells; *artificial intelligence*, which has used circuit design as a test-bed for general purpose problem solving techniques; *simulation-based optimization*, which attempts to improve circuit performance by using detailed circuit simulation within a general purpose optimization framework; and, *equation-based synthesis*, which uses designer-supplied expertise to achieve circuit performance specifications by determining cell topology, sizing and biasing. These approaches are important because they underlie our own work and have substantially influenced its evolution.

2.1 Layout-Based Design Automation

The layout-based approach to analog design automation is influenced by the gate array and digital standard cell technology of the early 80's [Kuh 90]. The strategy is to propagate physical manufacturing constraints to the circuit design process, making it quicker and easier to find a workable solution by limiting the designable space. Unfortunately, limiting the space of reachable designs means that the best solution to a specific problem will often be unattainable with layout-based design. The layout-based approach defines a range of ease of design/performance trade-offs that includes: standard cells [Pletersek 86, Stone 84, Lineback 87], field-programmable analog arrays[Lee 91], transistor arrays [King 85, Serhan 85, Pletersek 85, Pickerrell 83], and fixed topology module generators [Hobekotte 87, Yaghutiel 86, Allen 83, Degrauwe 84a].

At one end of this spectrum are mixed-signal standard cells. The strategy is to adapt semi-custom design techniques—the bulwark of traditional digital ASIC design—to mixed-signal applications. By augmenting hand-crafted digital standard cell libraries with analog cells, semi-custom mixed-signal ASICs can be designed by simply interconnecting these pre-designed circuit building blocks. But analog cells typically have a large number of performance constraints, making it difficult to cover adequately the space of designable cells. As a result, ASIC vendors typically offer cells that are as generally applicable as possible. However, systems with analog circuitry invariably require a few unique circuits that are very application specific. It is very unlikely that all the cells needed, including these unique ones, are available in a single vendor library. Attempts to "stretch" existing library cells to these unique requirements lead inevitably to poor overall performance. As a result, most mixed-signal ASICs designed with standard cells have good digital performance, but either employ analog cells that are not well suited for their final use—resulting in poor overall performance—or resort to time-intensive full-custom manual design.

At the other end of the spectrum lie fixed topology module generators. Here, the topology is completely fixed and encoded in the module generator. The *relative* geometry of the cell is also fixed. The cell is parameterizable in that the device widths and lengths can be altered somewhat, although devices maintain their relative position, allowing a single cell to achieve one of a few available performance trade-offs. This provides more flexibility than a library of completely fixed cells, but the range of circuit performance achievable is still extremely limited when compared to fully custom sizing. The module generators also require substantial effort to create. To justify this effort, a particular type of cell with a somewhat limited range of performance specifications would need to be required in a large number of designs. Further, the methodology more adequately addresses the cell layout problem, which is to generate final geometry given device dimensions, and not the cell synthesis problem, which is to provide a means to map the performance parameters to device dimensions.

In summary, if a library or module generator is available, layout-based techniques provide good automation because they are a fairly rapid path to silicon. However, these techniques do not have the breadth to push the limits of high-performance design, do not provide automation when there is no existing library or module generator, and do not have the scope to cover sufficiently the parts of the analog design flow in Figure 1.1 that are the central focus of this book.

2.2 Artificial Intelligence Approaches to Design Automation

Researchers into artificial intelligence have occasionally used the analog design domain as a test-bed for AI methods [Williams 84, Sussman 75, Roylance 80, Roylance 75] because it presents well defined but difficult design problems. These efforts have brought to bear such techniques as reactive expert systems [Trontelj 90], machine learning [Huhns 88], constraint propagation [Dekleer 80], fuzzy logic [Hashizume 89], and qualitative reasoning [Dekleer 85]. Unfortunately, much of this work focuses on the study of AI itself, rather than on creating tools that would be useful in an industrial circuit design environment.

Despite this, AI research on automating analog circuit design has had an important influence on some more practically oriented analog synthesis tools. CAD researchers have applied the general problem solving techniques developed by AI researchers to the analog synthesis problem. In general, these have focused on obtaining results that are more usable in the industrial environment and have achieved some success. In particular, tools such as OASYS [Harjani 89], incorporated AI planning and search techniques into a framework created specifically for analog circuit synthesis. More recently, ISAID [Toumazou 90, Makris 92, Makris 95] couples qualitative reasoning with equation-based analog circuit synthesis. Since these systems can also be categorized as equation-based synthesis tools, more detailed discussion of both tools is deferred until Section 2.4.

2.3 Simulation-Based Optimization

The third technique for analog cell design automation we discuss is simulation-based optimization. Optimization techniques have been a part of analog design for at least 40 years [Aaron 56]. With the advent of digital circuit simulation in the 60's [Nagel

71, Mathers 67, Ecap 65], substantial effort was expended on using computers to optimize circuit design [Rohrer 67, Temes 67]. These techniques were rediscovered when faster computers and improved simulators made them more practical for research [Nye 88, Shyu 88, Hocevar 90] and are now making their way into industrial CAD systems. Starting with a good set of initial circuit component values, these optimizers can adjust several of these values automatically, thereby more closely matching the circuit performance to a desired function, while satisfying constraints on the component values.

The goals of these optimization tools are not identical to those of synthesis tools because synthesis tools create a working circuit from a set of specifications, while optimization tools improve a circuit design that already works. However, several important ideas that originated in the area of analog circuit optimization are closely related to the work described in this book, so we will explore this area in greater detail.

2.3.1 Formulating Analog Design As A Numerical Optimization Problem

Although these circuit optimization tools do not perform synthesis, the formulation of the nominal circuit design problem as a general optimization problem is also a starting point for many synthesis tools (including ASTRX/OBLX) and worth describing in detail. To begin, we make the following definitions:

Def 1: Let $\underline{x}$ be the set of independent variables—physical geometries of semi-conductor devices or component values of passive circuit components—we wish to change to improve circuit performance. And n be the number of elements in $\underline{x}$. Then $\underline{x} \in \Re^n$. And we use x_i to denote the ith independent variable, such that $i \in [1, n]$.

Def 2: Let $\underline{f}(\underline{x})$ be a set of k objective functions. Again, $f_i(\underline{x})$ denotes the ith objective function, such that $i \in [1, k]$ and $f_i(\underline{x}) : \Re^n \to \Re$. These functions may include performance specifications that are common to most circuits, like power, or they may be circuit-specific, like the least square difference between the present transfer function and a desired one.

Further, real circuit design problems are typically subject to hard constraints, such as performance specifications not under a designer's control. This leads to the addition of inequality constraints, defined as:

Def 3: Let $\underline{g}(\underline{x})$ be a set of l constraint functions. Again, $g_i(\underline{x})$ denotes the ith constraint function, such that $i \in [1, l]$ and $g_i(\underline{x}) : \Re^n \to \Re$. These functions are formulated such that the constraints are satisfied when $\underline{g}(\underline{x}) \leq 0$.

The optimization task is to minimize the functions in $\underline{f}(\underline{x})$ while maintaining the constraints $\underline{g}(\underline{x}) \leq 0$. One thorny issue remains: we must define what exactly we mean by minimizing a *set* of functions. Typically, we do not wish to minimize each of the functions individually. Instead, one approach (the one used in our own work) is to minimize the weighted sum of the functions:

$$\underset{\underline{x}}{\text{minimize}} \sum_{i=1}^{k} w_i \cdot f_i(\underline{x}) \qquad \text{s.t.} \quad \underline{g}(\underline{x}) \leq 0 \tag{2.1}$$

where w_i is a scalar that normalizes the magnitude and balances the relative importance of $f_i(\underline{x})$. (For an alternative method see [Lightner 81].) We refer to Equation 2.1 as a *multi-objective non-linear constrained optimization problem.*

Although mapping the problem into a general optimization formulation is quite straightforward, solving the resulting problem is not so. In general, any implementation of this optimization strategy consists of two modules (see Figure 2.1): the optimization engine and the simulator. Unfortunately, practical implementations of these modules that together form a general, reliable, efficient and accurate system have proven elusive.

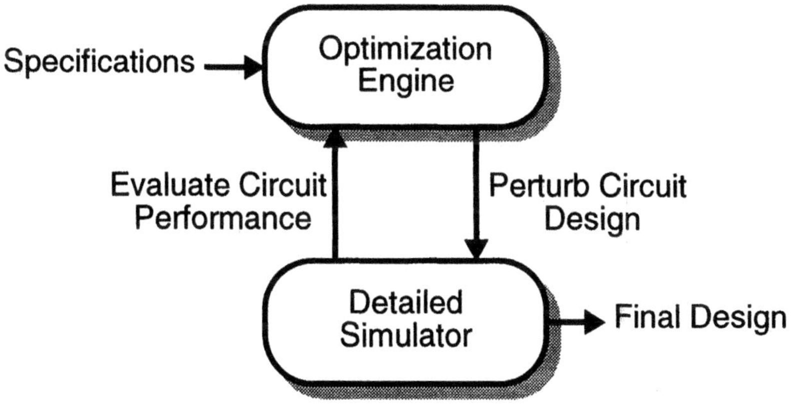

FIGURE 2.1 Implementing Circuit Design Via Optimization

2.3.2 Case Study: DELIGHT.SPICE

A representative example of a simulation-based analog circuit optimization system is DELIGHT.SPICE [Nye 84, Nye 88], developed at the University of California at Berkeley. It is an interactive tool designed to help an experienced designer explore trade-offs between performance objectives for circuits whose design is nearly complete.

The general structure of the tool is that shown in Figure 2.1. As its name implies, it uses an optimization engine called DELIGHT [Nye 81] wrapped around a version of the detailed circuit simulator SPICE [Nagel 71, Nagel 75] for performance prediction. This provides the generality and accuracy that come with a detailed circuit simulator but at the expense of very long run-times. SPICE is also notorious for divergence problems in its D.C. solution algorithm, a problem that can be fatal because the optimizer cannot recover when SPICE cannot evaluate a particular circuit configuration.

DELIGHT employs an aggressive optimization algorithm called the *method of feasible directions*. Following the terminology of Section 2.3.1, at each iteration of the algorithm, the optimizer perturbs $\underline{x}$ by moving in the direction of $\underline{h}$, where

$$\underline{h} = -\text{Nearest}\{\nabla \underline{f}_{\varepsilon a}(\underline{x}) \cup \nabla \underline{g}_{\varepsilon a}(\underline{x})\} \tag{2.2}$$

and Nearest $\{\underline{p}\}$ designates the nearest point to the origin in the convex hull of a set $\underline{p}$. Thus, the method of feasible directions (like many gradient-based optimization methods) attempts to find the point where the gradients of functions being optimized are zero. The major difference is the introduction of, $\underline{f}_{\varepsilon a}(\underline{x})$ and $\underline{g}_{\varepsilon a}(\underline{x})$, called the ε-*active sets*, which are subsets of $\underline{f}(\underline{x})$ and $\underline{g}(\underline{x})$ respectively. Informally, $\underline{f}_{\varepsilon a}(\underline{x})$ contains those functions whose present value is within some ε of $\max_{i \in k} f_i(\underline{x})$ (*i.e.*, close to the worst) and $\underline{g}_{\varepsilon a}(\underline{x})$ contains the constraint functions that are within some ε of being violated. This is an "aggressive" technique because at each iteration it optimizes only the worst functions.

Despite this aggressive strategy, circuit optimization with DELIGHT.SPICE is slow. Evaluating the functions and derivatives for one $\underline{x}$ requires about 130 CPU seconds[1] and an optimization takes about 2 hours for a moderately sized analog cell[2]. Since the

1. In 1988. Doubling CPU performance every 18 months, one would need to divide these values by about 12. Regardless, this is still slow.

tool was intended to require designer interaction, run-times this large are on the threshold of unacceptability. Because a less aggressive optimization strategy would greatly increase the run-time, interactivity was clearly a deciding factor in the choice of the method of feasible directions. The unfortunate consequence of this choice is that in practice, DELIGHT.SPICE can diverge without an excellent initial circuit as a starting point [Nye 84]. As a result, it is not very effective when used to make more than minor changes in circuit component values. For example, the component values of the circuits optimized in [Nye 88] typically changed by less than 25%, whereas a reasonable range for these values might encompass several orders of magnitude. Because optimization requires a functional design as a starting point, it means that optimization tools do not solve the synthesis problem, which by definition creates a functional design from specifications.

2.3.3 From Optimization to Synthesis

DELIGHT.SPICE and most previously proposed simulation-based circuit optimizers in general are more appropriately suited to final tuning of circuits, rather than performing the bulk of the design work. This is more the result of practical rather than theoretical considerations. In theory, one can consider the differences between optimization and design to be merely matters of degree, which we enumerate as:

1. An optimizer starts with a manually designed circuit that is "close" to correct, while a synthesis tool starts from completely unknown component values.
2. An optimizer adjusts a few of the circuit component values, while a synthesis tool must be able to derive all of them.

These may seem to be subtle distinctions, but in the past 15 years they have proved insurmountable. The key hurdles that have not been overcome to make this transition from *optimization* to *synthesis* are:

- **Efficiency:** Because SPICE-like simulators take so long to evaluate the performance of a single circuit design, the search mechanism must invoke the simulator as infrequently as possible. This can be seen in DELIGHT.SPICE, where the method of feasible directions is used because it requires few calls to SPICE. Within this framework it is imperative to reduce the complexity of the problem however possible, implying that the designer must spend substantial effort study-

2. A 28 transistor opamp with 6 constraint functions and 1 objective function. 20 circuit component values were varied. Only D.C. and A.C. analyses were required. Run-times were not given for optimizations requiring transient analyses, although the authors indicate that they would be substantially longer.

ing the circuit before DELIGHT.SPICE can be used. In fact, the authors of DELIGHT.SPICE concede that "only circuit parameters that have considerable effect on the design objectives and specifications should be considered,"[Nye 88] since including all the circuit parameters would be impractical. These requirements are not reasonable for synthesis, where the bulk of the design work should be performed by the synthesis tool. Although computation speeds continue to increase, we are likely many years away from workstations that are capable of running SPICE-like simulators fast enough to make their use in synthesis anything but painful.

- **Starting Point:** Tools such as DELIGHT.SPICE must be primed with a good initial circuit design. Without such a good starting point, an optimization may not converge or may converge to a local minimum significantly worse than the circuit's best capabilities [Nye 84]. In circuit synthesis, an aggressive gradient-based local optimizer is not practical both because the search space contains many local but non-global minima [Temes 67, Nye 84, Harvey 92] and because the user of a synthesis tool can not always be expected to provide a good initial circuit design.
- **Robustness:** Many SPICE-like simulators also suffer from convergence problems when solving for a circuit's initial D.C. operating point. These problems are exacerbated in an optimization environment where intermediate circuit designs in the optimization process may be essentially random and behave very strangely, making them more difficult to simulate. Aggressive gradient optimization techniques are not robust in the face of cost surfaces with gaps, and thus fail when the simulator fails to find a D.C. operating point for a given circuit configuration. Recently, new algorithms have been developed that can significantly improve the D.C. convergence capabilities of simulators [Trajkovic 91]. However, these algorithms also greatly increase the simulation time.

Although they are accurate and work on high-performance circuits, overall these optimization techniques do not adequately address our goals for synthesis because they do not automate enough of the design process. Nevertheless, the accuracy and breadth of optimization are desirable characteristics, and we will borrow ideas from this field to obtain these characteristics for our synthesis formulation.

2.4 Equation-Based Synthesis

Finally, we examine systems that employ the fourth technique for analog cell-level circuit design automation. These systems are called equation-based synthesis tools because they rely on a core of equations and can generate a sized circuit schematic to meet performance specifications. The central sizing algorithm for equation-based synthesis is, in many respects, much like simulation-based optimization. As shown in Figure 2.2, equation-based synthesis also employs an iterative design strategy. With each pass through the design loop, a search mechanism perturbs $\underline{x}$ (the element values, transistor dimensions and other variable aspects of the circuit) in an attempt to arrive at a design that minimizes $\underline{f}(\underline{x})$ and $\underline{g}(\underline{x})$ (more closely meeting performance specifications and other objectives). Performance equations are used to evaluate the new circuit design, determine how well it meets its specifications and provide feedback to the search mechanism to guide its progress. Two significant advantages of equation-based search over simulation-based methods are the increased speed of the performance evaluation and the additional flexibility in search strategy that can be employed to take advantage of circuit-specific knowledge. However, these advantages come at the expense of decreased accuracy and the prohibitive first-time design effort required to supply performance equations for each new circuit. Nevertheless, with the exception of our own work, all recent approaches to analog circuit synthesis fall into the equation-based category.

A fair number of researchers have explored variations on the equation-based synthesis theme [Trontelj 90, Toumazou 90, Sheu 88, Koh 90, Knapp 86, Klinke 92, Harvey 92, Gielen 90, Fung 88, Elturky 89, Degrauwe 87a, Degrauwe 89, Bowman 85, Berkcan 88]. We begin the remainder of this section with a generalized framework for equation-based synthesis that will allow us to categorize tools based on this paradigm. We then present case studies of a representative cross-section of these tools, focusing on their evolutionary progress. Finally, we describe comparison

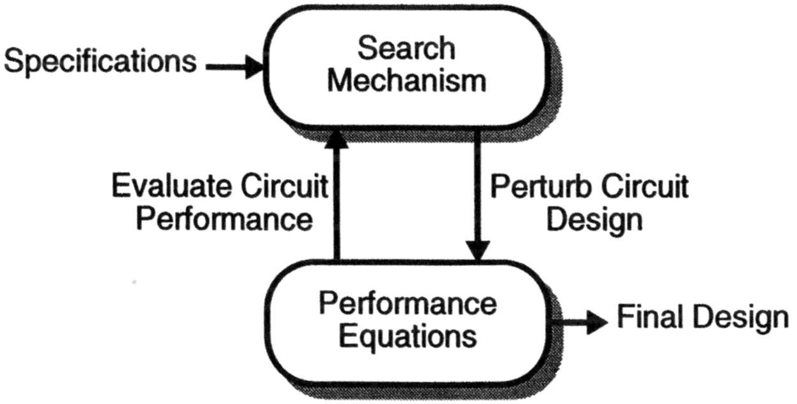

FIGURE 2.2 Search Process Used in Equation-Based Analog Synthesis Tools

criteria for equation-based synthesis tools and compare our sample systems based on these criteria.

2.4.1 A Generalized Model for An Equation-Based Synthesis System

Figure 2.3 shows a generalized model for equation-based synthesis tools. This flow automates the topology design and sizing and biasing tasks in Figure 1.2 but does not address variation tolerance. In this model, the path from specifications to a sized circuit schematic comprises these steps:

1. **Select Topology.** Selecting an appropriate interconnection of non-linear devices and linear circuit components is the first synthesis step. This task is often performed manually because effective topological *design* requires a depth of understanding and a degree of creativity that has proven difficult to automate. However, many equation-based tools incorporate automatic topology *selection*, which attempts to determine which of a set of *pre-designed* topologies is most appropriate for the problem at hand.
2. **Generate Analysis Equations.** Given a topology, analytic expressions must be derived for $\underline{f}(\underline{x})$ and $\underline{g}(\underline{x})$, which map the circuit variables to performance metrics of interest to the designer. These performance equations will be required to choose the device sizes (see step 3). Further, the equations must be massaged into a form that makes them amenable to solution via the methods selected for steps 3 and 4. This equation generation process must be repeated for each new topology because the results are topology-specific. This is typically the step in the synthesis process that requires the most human intervention, since there are no tools that completely automate the process.
3. **Sizing and Biasing.** Here the circuit is actually designed. Given the equations derived in step 2, a search or optimization technique is employed to find a set of values for $\underline{x}$ that satisfies the given performance criteria. Because the equations can be ordered, partially inverted, or forced into a pre-determined form, this search process is often heuristic, yielding further speed improvements over simulation-based methods.

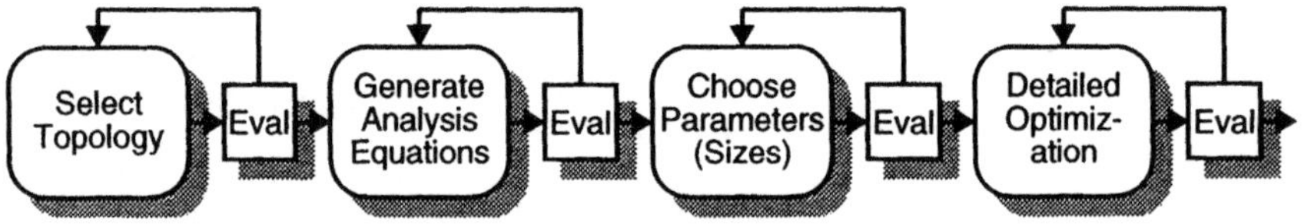

FIGURE 2.3 A Generalized Model for Equation-Based Synthesis

4. **Detailed Optimization.** Because of the approximate nature of the manually derived equations, a post-design optimization phase is often incorporated into equation-based tools. Here, the speed/accuracy trade-off is shifted by using more accurate non-linear device models and/or more general circuit simulation techniques. A complete simulation-based optimizer has even been used for this step [Onodera 90].

In practice, these steps are not always distinct. For example, by structuring the problem hierarchically it is possible to interleave topology selection and sizing.

Since we have explicitly decided not to address topology selection, we are most interested in steps 2, 3, and 4. In the remainder of this section, we discuss several case studies, concentrating specifically on the sizing, biasing and equation generation techniques employed by the tool. Comparisons and critical analyses of these tools is presented after the tools are described.

2.4.2 Case Study: OPASYN

OPASYN [Koh 87, Koh 90] was created at the University of California at Berkeley. It represents the logical first step from simulation-based optimization to equation-based synthesis. Ignoring an initial topology selection phase, the basic architecture of the tool (see Figure 2.4) is essentially the same as the simulation-based architecture of Figure 2.1. The significant change is the replacement of the detailed circuit simulator with hand-crafted equations.

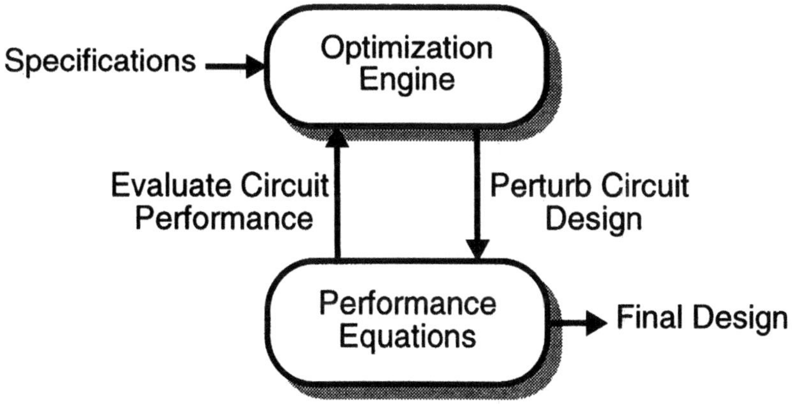

FIGURE 2.4 OPASYN

Sizing and Biasing

OPASYN employs constrained optimization techniques (see Section 2.3.1) to perform sizing and biasing to meet performance specifications. The optimization engine used is a simple steepest descent algorithm that follows gradients to the nearest local minima [Luenberger 84, Fletcher 90]. This mechanism is not as aggressive as the optimizer in DELIGHT.SPICE, but the tool runs substantially faster because of the analytic expressions used for performance prediction.

OPASYN relies on several techniques to aid in converging to a reasonable solution. First, the creators of the hand-crafted circuit performance evaluation equations define very tight bounds on the allowable values for variables, using designer expertise to determine each variable's "useful range in practical applications"[Koh 90]. Second, at the start of each optimization, this tightly bounded search space is gridded and several starting points randomly picked as starting points for optimization. "OPASYN reports all the different solutions (if any) that meet the specifications to the user who selects one for further processing." [Koh 90] This gridded multi-start approach to optimization attempts to address the local minima problem by providing a selection of local minima from which to choose, increasing the odds of finding a global minimum. Unfortunately, the number of required starting points scales exponentially with the dimensionality of the problem, making the approach impractical for even a modest number of circuit component variables.

Equation Generation

In OPASYN, the user is required to provide both the equations that map the circuit component values to performance characteristics and the analytical derivatives of those equations. To make this onerous task possible, the designers of OPASYN rely on "first-order circuit analysis techniques and topology specific approximations"[Koh 90]. These expressions differ by as much as 200 percent from simulation results, forcing the use of "fitting parameters" to make OPASYN's performance predictions more closely match those of a simulator. In practice, these fitting parameters need to be tuned by running a large number of circuit simulations for each new circuit and each set of process parameters. To guide the optimization process, analytical derivatives are required for each of the equations with respect to each of the independent variables. A great deal of expert design effort is also required to reduce the number of independent variables, both to reduce the number of derivatives and to reduce the difficulty of the optimization task. Reducing the dimensionality of the optimization problem is especially important when using a gridded random multi-start optimization method, as discussed in the previous section. The largest OPASYN example in [Koh 90] contained 60 transistors, but only 7 independent variables, substantially fewer than one would expect for fully automatic design. The design expertise and effort required to generate the equations for a single circuit is formidable. For an expert designer, who had already designed the same opamp manually, deriving and entering the equations for a circuit with 24 active devices took about "two weeks". Because "OPASYN relies on analytic models specially developed for each of the opamp circuit topologies"[Koh 90] this process must be repeated for each circuit OPASYN can design.

OPASYN was important because it used equations within the well-known optimization formulation for analog circuit design. However, it suffered both because of the simplicity of its performance equations and because of the difficulty in deriving them.

2.4.3 Case Study: OASYS

OASYS [Harjani 87, Harjani 88, Harjani 89] represents a substantial departure from the optimization style of analog synthesis. Instead, OASYS presents a fairly general software framework for *knowledge-based synthesis*, customizing it for analog circuit design. Knowledge-based synthesis draws heavily from expert system techniques in artificial intelligence [Davis 82, Brown 85, Rychener 85]. OASYS seeks to formalize the mechanisms used by human designers and capture this expertise in repeatable form. To accomplish this, OASYS relies on three key organizational principles:

1. *Hierarchy:* The decomposition of a large problem into a small number of simpler problems. In OASYS, circuits are decomposed into smaller sub-circuits, and these sub-circuits are themselves successively decomposed. Eventually, transistors and other fundamental components are designed explicitly and the completed circuit re-assembled.
2. *Style Selection:* The selection of an interconnection of sub-circuits (a topology where each component may itself be decomposed hierarchically) to best satisfy a particular given set of performance specifications.
3. *Translation:* The mathematical mapping of performance specifications for a particular circuit in a design hierarchy into sets of specifications for each of its component sub-circuits.

Sizing and Biasing

In OASYS, topological design, sizing, and biasing are interleaved. A circuit is synthesized by repeated applications of *style selection* (topological design) and *translation* (sizing) at each stage in a hierarchical description of a circuit (see Figure 2.5). Although more sophisticated techniques are possible, style selection is typically performed by generating designs using all available styles and selecting the one that best matches the specifications. The translation process proceeds as a series of *plan steps.* This planning paradigm was originally intended to follow the reactive planning style of AI systems [Schank 77]. However, in practice, the plans are a fixed set of steps created by an experienced analog designer who understands the OASYS translation methodology. Similar to the techniques used by human designers, plan steps perform three sorts of activities:

1. *Heuristics:* Educated estimates of circuit parameters based on incomplete information. For example, device parasitics are needed to calculate performance. However, parasitics cannot be calculated until the device dimensions are known, *i.e.,* when the design process is complete. To break this dependency loop, parasitics are estimated.
2. *Computation:* Calculation of circuit parameters or lower level block specifications using first-order circuit models.
3. *Refinement:* Invocation of the entire style-selection/translation process for a component of the block presently under design.

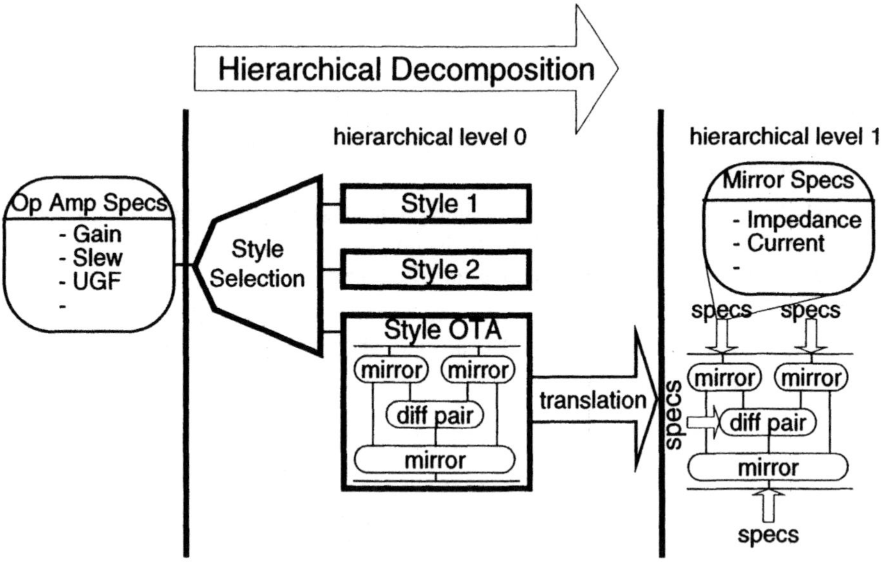

FIGURE 2.5 OASYS

Because translation often involves estimates based on incomplete data, OASYS provides a sophisticated backtracking mechanism to allow portions of a design plan to be repeated when better estimates of the estimated parameters are available. This backtracking mechanism provides OASYS with a reactive programming capability similar to rule-based production systems. Repeated translation/backtracking iterations also allow rudimentary optimization as the estimates and actual calculations converge.

Equation Generation

The hierarchical decomposition of circuits simplifies the task of equation generation somewhat. In effect, hierarchy forces the behavior of a sub-circuit to be macro-modelled using its port behavior. Thus, only equations for the interactions of a few sub-circuits must be derived for each hierarchical level. This is a significant advantage of the hierarchical organization introduced by OASYS. However, as discussed in Section 1.3, in many cases important secondary effects cannot be hidden when designing high-performance analog circuits. As a result, as more information is hidden within the design hierarchy, the accuracy of the synthesis model erodes. Despite this drawback, OASYS' hierarchical approach is employed by most of the analog synthesis tools that follow it.

OASYS drew ideas from AI techniques and several more primitive knowledge-based analog circuit synthesis systems [Elturky 86]. After its introduction, many new and existing tools drew heavily from the ideas used to create the OASYS framework, *e.g.*, [Sheu 88, Berkcan 88].

2.4.4 Case Study: IDAC

IDAC [Degrauwe 84a, Degrauwe 87a, Degrauwe 87, Degrauwe 89] was originally developed at the Centre Suisse d'Electronique et de Microtechnique (CSEM) and was then commercialized by Mentor Graphics Corporation.

IDAC was one of the first analog synthesis tools introduced after the resurgence of interest in analog design automation in the early 80's. It started as an *ad hoc* collection of software modules for designing circuits from a library of fixed topologies [Degrauwe 84a], but evolved continually since its introduction and is arguably the most mature analog synthesis tool presently available.

Sizing and Biasing

IDAC employs a two-step sizing algorithm. As shown in Figure 2.6, an initial rough sizing phase is followed by a more detailed numerical optimization phase. For the initial rough sizing, simplified analysis equations are inverted to solve for device dimensions given performance specifications. In the optimization phase, gradient-based optimization techniques and a more detailed analytic circuit model are employed. Originally, the architecture used by IDAC to organize its analog design knowledge was somewhat *ad hoc*. More recent versions of IDAC [Degrauwe 89] store analog design expertise in a structured form much like that used in OASYS.

Equation Generation

The most conspicuous innovation introduced by IDAC is the use of symbolic simulation [Seda 92, Seda 88] to aid in the creation of the analog design expertise. Symbolic simulation uses simulation algorithms to evaluate circuit performance, but performs the algebra symbolically. Thus, the final result is an expression. These tools were introduced primarily as aids to expert designers. It appears that designers of these early tools did not see the full potential of symbolic simulation in fully automatic synthesis, and the exploitation of symbolic simulation as a pivotal part of the overall synthesis process was left to the ARIADNE system (Section 2.4.5).

More recently, the researchers on the IDAC project have introduced a special purpose language [Jongsma 91] for describing their analog circuit expertise. At present, this language is capable of describing only a subset of the circuits IDAC has in its database. This language gives the user the capability of augmenting that database with some simple circuits of her own. However, an intimate understanding of how IDAC works is required to add additional analog expertise to the system.

Perhaps the strongest part of the IDAC system is the quality of the analog design expertise stored within it. IDAC has had a team of designers refining and adding to its

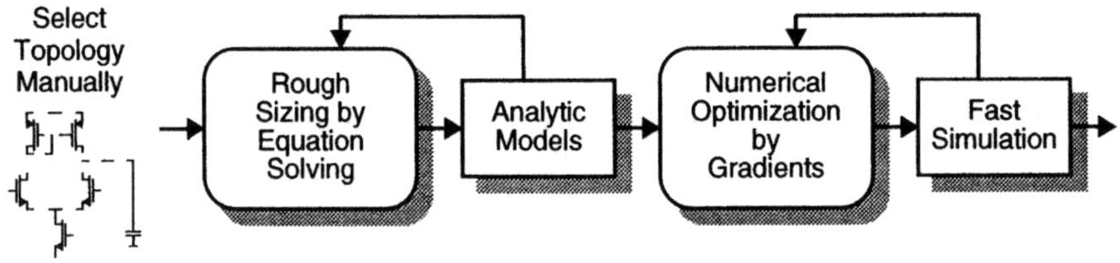

FIGURE 2.6 IDAC

circuit library since the tool's creation. As a result, the system can synthesize circuits from 15 different classes. However, the range of applications for analog circuits in general is far broader than the IDAC library. Moreover, proprietary circuit designs are often the lifeblood of a company. These two factors limit the widespread acceptance of a system with a fixed repertoire of designable circuits.

The directions IDAC will continue to evolve are unclear. Its designers introduced a speadsheet driven system for creating analog circuit design plans [Henderson 93]. The structure of plans are essentially identical to the structure used by OASYS, and the tool is called PlanMaker. PlanMaker is substantially more open than the IDAC-based tools because none of the circuit specific knowledge is built into PlanMaker itself. However, this style of planning does not appear to be wholly compatible with the circuit knowledge stored in IDAC. Since there is a substantial investment of effort in the knowledge already stored in IDAC, it seems unlikely that IDAC will be completely abandoned in favor of the newer spreadsheet-based approach.

2.4.5 Case Study: ARIADNE

ARIADNE [Swings 91, Gielen 89, Gielen 91, Wambacq 91, Swings 91a, Gielen 90a, Gielen 90] is a suite of analog CAD tools from Katholieke Universiteit at Leuven, Belgium. It marked the beginning of a second generation of equation-based analog synthesis tools that built upon the ideas of OPASYN, OASYS and IDAC.

Sizing and Biasing

ARIADNE sizes and biases the circuits it designs using an optimization tool called OPTIMAN, which is similar to other optimization-based approaches described previously (see Figure 2.7). One important improvement is the use of simulated annealing as the optimization algorithm. (See Section 4.1 for a more detailed description of simulated annealing.) However, we believe that OPTIMAN does not make full use of the potential of annealing, which is likely an important reason why the circuit synthesis results published for ARIADNE are for easily designable circuits. In other applications, simulated annealing provides freedom from starting point problems and robust optimization over complex cost surfaces with many local minima. However, the price for this optimization robustness is a very large number of evaluations of the cost function. Unfortunately, in the ARIADNE system, a single evaluation of a circuit requires about 30 seconds [Gielen 90a]. Because we can expect to evaluate a circuit tens of thousands of times to effectively anneal a typical synthesis problem, 30 seconds is about an order of magnitude too slow for a single evaluation with an annealing-based optimizer. To compensate for this, OPTIMAN uses annealing parameters that make as few circuit evaluations as possible, leading to an optimization tool that behaves much like a downhill optimizer without gradients. As a result, the few circuit design results published for ARIADNE are textbook circuits with fairly low performance requirements.

Equation Generation

The key innovation in the circuit synthesis approach of ARIADNE is the reliance upon symbolic simulation as an integral part of the synthesis process. Using the symbolic simulator ISAAC [Gielen 89, Gielen 91], ARIDANE attempts to address the largest problem of the knowledge-based synthesis approach typified by OASYS: the difficulty of deriving new circuit design knowledge. ISAAC uses symbolic techniques to derive analytic small-signal transfer functions for arbitrary circuit topologies. Active devices are replaced with small-signal elements. After the addition of large-signal equations provided by a designer, the complete set of equations that map circuit component values to circuit performance is passed to the equation manipulation tool DONALD. The role of DONALD is to determine an order in which the equations can be solved (sometimes involving simultaneous solution when there is a dependency loop) to map from a given basis set of independent circuit component values to circuit performance.

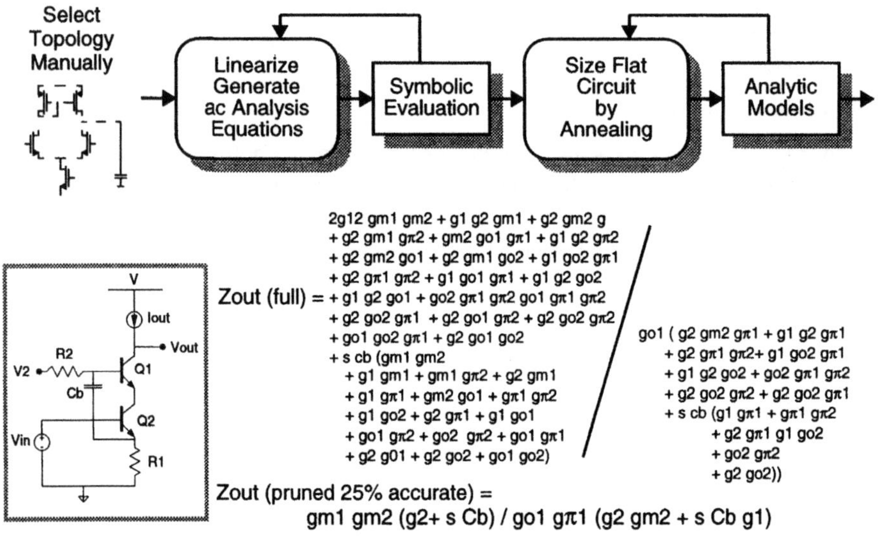

FIGURE 2.7 ARIADNE

In theory, the use of ISAAC could be very powerful. Symbolic simulation generates analytical transfer functions for small linear or linearized circuits. For many important class of circuits, such as operational amplifiers, most of the important specifications are linear in nature and are amenable to this kind of analysis. Moreover, fairly accurate equations for many of the remaining specifications can be derived by inspection, whereas equations for the linear specifications are much more involved. However, symbolic simulation has yet to overcome substantial technical obstacles before it can fully automate performance equation derivation for large, high-performance circuits. The principal problem is that the number of terms in an analytic transfer function grows exponentially with circuit complexity. For example, the output impedance, Zout(full), of the common-emitter stage shown in Figure 2.7 contains nearly 50 terms, while experiments with a 9 transistor two stage opamp (see Section on page 144 for a description of this circuit) yield a transfer function with over 53,000 terms. To help combat this problem, ISAAC does include several algorithms for pruning these terms (see the pruned expression in Figure 2.7). However, these prune the terms based on magnitude and require a numeric point near which to make the approximation valid. Both of these requirements are problematic. First, phase information is very important for analog design and pole matching techniques lead to small terms in the transfer function which have very important effects on circuit performance. Second, for synthesis, a starting point around which to prune is simply not available. If a guess is made, there is a high risk that the region in which the approximation will be valid will soon be left behind. The net result is that ISAAC's creators are now focusing on the application of symbolic synthesis to instructing or guiding human designers rather than completely automating the circuit design process [Gielen 92]. However, work is continuing with other symbolic synthesis tools [Seda 92, Fernandez 92] and algorithms for more accurate simplification of large symbolic formulae are beginning to emerge [Fernandez 92, Hsu 94, Wambacq 95]. These new algorithms consider ranges of potential variation in circuit parameters when simplifying the symbolic expressions. If these algorithms are shown to be effective for practical circuits, symbolic simulation methods may play an even larger role in future synthesis tools.

2.4.6 Case Study: STAIC

STAIC [Harvey 92] is a more recent synthesis system from the University of Waterloo. Although most of the ideas in STAIC are not new, the system is noteworthy for its effective combination of techniques from previous tools.

Sizing and Biasing

STAIC uses a two stage approach to sizing and biasing, but both steps are optimization based. The first step uses simple circuit models and scans the designable space using a fairly coarse grid. Using the best results from this first scan as a starting point, the second stage uses more detailed models and a conventional optimization technique to obtain a final solution. Although these optimization models use flattened numerical models of the circuit, the circuit knowledge in STAIC is stored hierarchically. Similar in intent to the DONALD tool in ARIADNE, a pre-processing phase in STAIC flattens the hierarchical description for each circuit topology to be optimized.

Equation Generation

The most innovative component of STAIC is a problem description language that defines the analog circuit specific knowledge operated upon by the system. STAIC evolved at roughly the same time as the addition of a language to IDAC, and it appears that the two tools evolved independently. STAIC takes advantage of the inherent flexibility of C++ by pre-defining some special classes and operators that describe circuit connectivity, etc., that become the STAIC circuit description language. Because C++ is the implementation language for STAIC, design expertise becomes an integral part of the system. The complexity of circuit descriptions in STAIC provides a fairly good benchmark of the effort required to add new circuits to equation-based tools. In [Harvey 92], three circuit examples are presented (the third is actually a simple variant of the second). In total, these three circuits required about 10000 lines of STAIC code. So, although STAIC is an open system in the sense that users can add their own circuits to it, the amount of effort to do so is not much less burdensome as that required for OASYS or OPASYN.

2.4.7 Case Study: ISAID

ISAID [Makris 92, Toumazou 90, Makris 95], created at the Imperial College in London, is another recent analog synthesis system that, in many ways, builds on its predecessors. However, it includes a unique use of qualitative reasoning techniques instead of optimization to adjust an initial circuit design to match simulation more closely.

Sizing and Biasing

ISAID also performs sizing and biasing in two passes. In the first pass, ISAID performs a conventional hierarchical equation-based synthesis that is almost identical to OASYS. However, once this initial design is completed, adjustments to the design to compensate for the simplicity of the hierarchical equation model are performed using an artificial intelligence technique called *qualitative reasoning*. Simplifying somewhat, qualitative reasoning uses the sign of the gradient to predict how a performance specification will be affected by a change in a given device size or bias. Using chains of such predictions, the circuit parameters can be adjusted to obtain the desired overall change in performance. This method can be more effective than optimization for such adjustments, because the adjustments can also perform topological changes if necessary, while conventional optimization techniques operate only on a flattened circuit.

Equation Generation

The system provides little help in equation generation. In fact, the problem may be exacerbated to some extent by the need to provide additional information to guide the qualitative reasoning process.

2.4.8 Case Study: Maulik

Our final case study of an equation-based synthesis tool is the recent work of Maulik at Carnegie Mellon [Maulik 91, Maulik 92, Maulik 92a, Carley 93].

Sizing and Biasing

The work of Maulik includes two important innovations for sizing and biasing: the relaxed D.C. formulation and simultaneous topology selection and sizing within an optimization framework. Simultaneous topology selection and sizing uses a mixed-integer non-linear programming method to solve for both the device parameters and the best topology. Essentially, competing topologies are assigned integer values, and the optimizer solves for the appropriate integers while determining devices sizes. Of more importance to the work in this book is the relaxed D.C. formulation. This formulation relaxes the requirement that each circuit visited during iterative optimization be D.C. correct. Just as a synthesis tool formulates optimization goals such as meeting gain or bandwidth constraints, Maulik formulates D.C. correctness as yet another goal to meet. The idea of combining D.C. and performance constraints in a single optimization problem is not new [Lipparini 82]; however, it has been controversial [Kundert 89, Rhyne 89]. In [Kundert 89], Kundert and Sangiovanni-Vincentelli argue that although optimizing performance and solving for the D.C. operating point can be formulated as a single optimization problem, this formulation results in a problem that is less efficient to solve and less numerically robust than an independent formulation, which solves for the D.C. operating point within each step of a performance optimization. Simulation-based optimizers are a typical example of this independent formulation. Other equation-based techniques avoid this problem entirely by analytically solving the D.C. biasing equations for the circuit in advance. But this is possible only because they use simple device models, which can be easily manipulated algebraically. To allow the use of device models of industrial complexity without the other numerical techniques in simulators, Maulik chooses the combined formulation, which we refer to as *relaxed D.C.* He formulates D.C. and performance goals as constraint functions, then uses sequential quadratic programming techniques (SQP) to solve the resulting constrained optimization problem. Despite the misgivings of Kundert and Sangiovanni-Vincentelli, this approach has proven successful in efficiently converging to accurate solutions. However, the SQP optimization technique still requires a good starting point, and Maulik's system still requires difficult circuit-level (but not device-level) equation extraction and simplification.

Equation Generation

Maulik provides no automatic support for equation generation.

2.5 Comparison of Previous Systems

In this chapter we have described in detail several synthesis tools that together represent the state of the art in nominal analog synthesis. Table 2.1 summarizes the important characteristics of the representative equation-based synthesis systems detailed in Section 2.4. In this section we compare each system with each other and with an ideal system based on five key characteristics of an analog synthesis tool: accuracy, run-time, preparatory effort, generality/complexity, and openness. We begin with Table 2.1 and then discuss these five characteristics in greater detail.

2.5.1 Accuracy

Accuracy was first introduced in Section 1.4 as one of the four metrics to evaluate analog synthesis tools. The best measure of accuracy is the discrepancy between a synthesis tool's internal performance prediction mechanisms and the actual circuit when fabricated and tested. Since fabrication results are unavailable for almost all the published synthesis results, we typically compare the tool's predictions with a detailed circuit simulator that uses realistic device models.

TABLE 2.1 Comparison of Synthesis Tools in Section 2.4

Characteristic	Synthesis Tool						
	OPASYN	OASYS	IDAC	ARIADNE	STAIC	ISAID	Maulik
1. Performance evaluation	Equations	Equations	Equations	Equations	Equations	Equations/ Qualitative Reasoning	Equations
2. Non-linear device models	Simplified equations	Simplified equations	Custom models	Simplified equations	Simplified equations	Simplified equations	Equations + BSIM
3. Worst-case accuracy	200%	25%	15%	10% (low-perf.)	24%	14%	24%
4. Search Method	Gridded, Steepest descent optim.	Plan steps with backtracking	Plan steps with post optim.	Simulated annealing	Coarse initial optim. + detailed final optim.	Qualitative reasoning + post optim.	Sequential quadratic programming
5. Synthesis time (approximate)	1 min. VAX 8800	5 sec. VAX 8800	"A Few seconds"	5 min. (CPU N/A)	3 min. MIPS 2000	Not reported	1 min. DEC 3100
6. Approximate effort to add new circuit	2 weeks for well understood circuit.	6 months, including circuit analysis.	4 - 45 designer-months	Not reported	3 circuits required 10000 lines of code.	Not reported.	6 months, including circuit analysis
7. Equations derived	Manually	Manually	Symbolic simulation + Manually	Symbolic simulation + Manually	Manually	Manually	Manually
8. How/Where equations are stored	Hard-coded	Hard-coded	Hard-coded + Design files	Not reported.	Language + Database	Not reported.	Hard-coded
9. Most complex circuit example	7 variables (60 device opamp)	19 variables (17 device opamp)	N/A (15 circuit types, incl. delta-sigma, from 5-30 devices)	14 variables (9 device opamp)	N/A (22 device opamp)	8 variables (13 device opamp)	39 variables 7 for topology selection. (19 devices in largest opamp)

The fundamental reason for inaccuracy in equation-based approaches is that they rely heavily on simplifications to both circuit equations and device models. The relationships between the device dimensions of transistors and the performance of an analog circuit are non-linear, extremely complex, and often poorly understood. As a result, designers and the people who add knowledge to synthesis tools must make simplifications to keep the equations they manipulate tractable. Here the experienced designer has an advantage because she knows which simplifications can and cannot be made for the specific problem at hand. Because these simplifications may be entirely different for the same circuit with a different set of specifications, the problem is much more difficult when writing general purpose equations that will become part of a synthesis tool. Thus, much more effort is required to write the synthesis knowledge for a given circuit than simply to design a single circuit for a specific task. This trade-off between effort put into the performance equations and accuracy of the results is best seen when comparing the worst case accuracy and approximate effort to add a new circuit (lines 3 and 6 in Table 2.1). For example, the authors of OPASYN reported that they spent 2 weeks coding circuit knowledge for their circuits and that this knowledge differed from simulation by as much as 200%, while for IDAC substantially more effort was expended but the results matched simulation much more closely. Thus the accuracy of results achieved with equation-based tools is proportional to the amount of effort invested in deriving the equations. Although this fact may seem quite obvious, effort is difficult to quantify and in the literature, tools are rarely compared on this basis.

The second key factor leading to inaccuracies in equation-based synthesis tools are the simplifications made in modelling non-linear device behavior. In some sense, this can be seen as a special case of simplifications to overall circuit performance equations; however, the two kinds of equations are typically treated independently. As can be seen in line 2 of Table 2.1, almost all the equation-based tools also rely on simplified equations for their non-linear device models. Device modelling is itself an extremely difficult task and the subject of much ongoing research. The need to support the latest modelling technology, deep submicron devices, and high-performance circuit design goals is fundamentally at odds with equation-based strategies that rely on these simplifications. However, replacing these simple models with models of complexity comparable to those in detailed simulators is *not* straightforward. These device models often contain several thousand lines of equations, but this is only one of the difficulties in using them within synthesis tools. The most important problem with using completely independent device models such as those in detailed simulators is that this would negate a critical simplification in the search for the appropriate circuit solution. A simulator, where complex device models must be supported, solves for the D.C. operating point for the circuit using a numerical technique such as Newton-Raphson. This technique is the cause of several practical problems in simulators. First, Newton-Raphson may require a large number of iterations to find the D.C. operating point and each iteration may require device models to be re-evaluated for every non-linear device in the circuit. This process often consumes a substantial fraction of the overall simulation time. Second, the Newton-Raphson algorithm itself has well known convergence problems [Fletcher 90], and these can be triggered by numerical instability in the non-linear device models. These problems can be avoided completely by using simplified equations and solving for the D.C. operating point analytically. However, this simplification can dramatically affect the accuracy of circuit performance prediction. As discussed in Section 2.4.8, Maulik takes the first steps toward incorporating industrial strength device models by directly using BSIM [Sheu 87] models to evaluate the device small-signal parameters. However, he stops short of a full solution because these BSIM specific small-signal parameters are then used in other performance equations, preventing the use of different device models.

2.5.2 Run-time

Run-time is the CPU time necessary to synthesize a working circuit given that equations for that circuit are already in the tool's library. Run-time is one of the two components of the automation metric introduced in Section 1.4. Run-times for the tools we are comparing are in line 5 of Table 2.1. Of course, these are difficult to compare directly because of the different underlying CPU speeds. Further, these times are approximate and were selected as representative of run-times reported.

Equation-based tools appear to design circuits quickly. But, the stated run-time of the tools can be very misleading as a measure of automation because it does not consider the time required to derive the circuit equations (see line 6 of Table 2.1). In all cases, this derivation time far outweighs the run-times. The usefulness of tools with quick run-times increases with the number of times a given circuit topology will be redesigned to different specifications. Clearly, a particular topology would need to be used a great deal to justify its codification in an equation-based synthesis tool.

2.5.3 Preparatory Effort

Preparatory effort, the other component of automation introduced in Section 1.4, is the measure of how much time/effort is required to use a synthesis tool when trying to design a circuit that is *not* already in the synthesis tool's library. Note that we are not really interested in user interfaces and other factors that are outside the scope of our work. Since most equation-based tools begin with only one or two circuits in their library, and most design houses want to use proprietary circuit topologies, the time/

effort required of a user to render a new circuit design in a form suitable for a synthesis tool to complete is one of the most important measures of a tool's usefulness in practical design. Although it is often difficult to measure accurately, we compare preparatory effort by comparing both circuit analysis (equation derivation) time and/or lines of input needed to describe the performance equations.

As stated before, equation-based tools appear to design circuits quickly, but the stated run-time of the tools is very misleading because it does not consider preparatory effort. As shown in lines 5 and 6 of Table 2.1, including this derivation time, equation-based synthesis tools require substantially more effort for a first-time circuit design than the already time-consuming manual design process. With time-to-market the overriding concern, it is unrealistic to expect designers to be able to afford the time needed to teach equation-based synthesis tools to do their high-performance cell design.

2.5.4 Generality/Complexity

Another important characteristic of a synthesis tool that is often overlooked is its ability to handle different circuit types and large design problems. These are aspects of the breadth metric introduced in Section 1.4. The most complex circuit example synthesized by each of the sample tools is reported in line 9 of Table 2.1. For all of these tools, the circuit sizes designed are quite modest (even for analog circuits) and the number of independent variables is typically under twenty. Also, with the exception of IDAC, all these tools design textbook linear operational amplifiers whose circuit equations are relatively straightforward to look up or derive. In the case of IDAC, more complex circuits are included, but these take up to four years to analyze and enter into the system. Because modelling circuit performance with a set of mathematical equations is so general, it can be argued that equation-based synthesis is inherently the most flexible technique. However, the difficulty of deriving equations and the difficulty of solving circuit problems with a large number of independent variables makes this flexibility difficult to exploit in practice.

2.5.5 Openness

The openness of a system can be defined simply as the ability of the user to edit and supplement the library of circuits the system can design. This information is included as line 8 in Table 2.1. In early synthesis systems, the circuit-specific analog design expertise was hard-coded as part of the system itself, making it almost impossible for anyone except the original author to augment or modify this expertise. More recently, systems typically provide some access to the expertise and store it independently of the core routines in the system. This access is essential because most industrial analog design techniques are proprietary and cannot be added to a system until after it is purchased. Even with the ability to edit this design expertise, the designer doing so must typically be an accomplished programmer. Often, she requires experience with the programming language used to implement the tool—which may be C, C++, Lisp or even a research language—and an intimate understanding of the internal architecture of the tool, in addition to analog design skills. This is a very uncommon combination of skills. This difficulty in being able to modify or add to the analog expertise in a synthesis tool is an key stumbling block in the acceptance of these research-oriented tools in the industrial community.

2.6 Where Equation-Based Tools Need to Improve

Tools that attempt to synthesize analog circuits have emerged only in the last 10-15 years so there is much room for improvement. However, any list of areas in which these tools as a class could and should improve in the near future will of course be quite subjective. With that statement as a disclaimer, the ways in which equation-based tools as a whole need most improvement can be discussed in terms of the same criteria used to compare them (see Section 2.5):

- **Accuracy:** Equation-based tools need to evolve towards the accuracy of detailed circuit simulators when predicting circuit performance.
- **Run-time:** Most equation-based tools run fast enough that users should have no complaints about run-time.
- **Preparatory Effort:** This is the biggest failing of equation-based tools. Currently, humans are required to describe the circuit to be designed in excruciating detail to the synthesis tool. The tool then simply replays this stored information when performing the design task. This must change.
- **Openness:** With the advent of languages that allow users to describe design expertise, most tools are open enough to allow their circuit library to be edited. Further work here should not be a priority.
- **Generality/Scalability:** The circuits previously published as synthesis results are too small to be of use in industry. Although this appears to be a failing in the design expertise rather than a fundamental problem with the tool itself, a tool that provided greater automation would make generating extensive examples easier. Researchers must show that larger, more complex circuits can be designed.

2.7 Summary

In this chapter we explored previous approaches to design automation for cell-level analog circuits. Although there have been layout-based, AI-based, and simulation-based approaches to the problem, the most practical and successful approach has been equation-based. The layout-based approaches are too inflexible to produce high-performance circuits, while AI-based approaches have been used only for circuit designs that are too simple to be of interest to industrial analog circuit designers. Simulation-based optimization has been more successful, but is too slow and unreliable to synthesize a circuit without an excellent initial design as a starting point. In this application, only the equation-based techniques have met with success. However, even equation-based techniques have not garnered much use in the industrial community. This lack of acceptance is due to many factors, but primarily caused by the prohibitive one-time effort required to derive the complex equations that form the core of any equation-based synthesis system. Today's deadline-driven industrial design community simply cannot afford the additional time to invest in deriving circuit synthesis expertise in the hopes that it will be re-used later. Instead, a new analog cell synthesis methodology is needed to show that analog synthesis is a viable alternative to manual design.

CHAPTER 3 A New Nominal Synthesis Strategy

In this chapter we present our strategy for a new analog synthesis system. We begin with a description of the goals of our work, present the five interrelated ideas with which we will achieve these goals, describe how these ideas have been formulated into a new synthesis architecture and conclude with an extensive example that shows how a real circuit problem is synthesized within this new architecture.

3.1 Goals

In Section 1.4 we introduced the goals of analog synthesis and defined some metrics to evaluate these goals. In Section 2.6, we described how existing analog circuit synthesis methodologies fall short of these goals. With our new nominal synthesis strategy we push beyond the accomplishments of existing approaches. However, realistically, we cannot hope to achieve progress will all our goals without making some trade-offs. In this section, we prioritize the areas in which we most wish to improve upon the prior approaches to analog synthesis, and enumerate the trade-offs that we are and are not willing to make to achieve these improvements.

1. **Automation.** Our primary goal is to reduce or eliminate the one-time effort required by analog designers to codify analog circuit design expertise that describes the circuits they wish the tools to synthesize. Because of this effort, it is

effective to use equation-based tools only for circuit topologies that will be re-used many times. We wish to make it desirable to use a synthesis tool even for a new circuit topology that will be used only once.

2. **Accuracy.** Our secondary goal is to improve the accuracy of synthesis tools so that their performance predictions more closely match simulation. As we discussed in Section 1.4, accuracy is the key enabler for high-performance circuits design because complete control of the synthesis process is in the single tool that is making the design decisions, alleviating the need for painful manual or automatic tuning after the synthesis tool has finished.
3. **Breadth.** Our final goal is to create a tool that can find practical application in industry. In terms of generality and complexity, this means that we must be able to handle a broad spectrum of circuits, not just a handful of circuits from a small library, and handle design problems with the complexity faced by industrial designers, not just the circuits seen in analog textbooks.

To achieve these goals we enumerate the long-term and short-term design-trade-offs we are willing to make. For the long term we are willing to accept:

- **Increased run-time.** There are two reasons why increased run-time is an acceptable trade-off. First, computer speeds continue to increase while the number of devices in an interesting analog circuit remains more or less unchanged, so techniques that seem slow today may not seem slow in the future. And more importantly, we believe a designer's time is more important than a computer's time. We are more than happy to trade several months of designer effort deriving circuit analysis equations for several hours of CPU time.

In the short term, we are also willing to accept:

- **No topological design.** We believe that sizing/biasing is the correct starting point for a new synthesis strategy. As long as the new strategy does not preclude topological design, a suitable mechanism can be added later. Exploratory work to confirm the viability of this extension was performed with encouraging results [Jones 96].
- **No operating range and manufacturability concerns.** This trade-off is very short term and only for clarity of explanation. After we complete the description of our nominal synthesis system in Chapter 7, the second part of this book describes our extensions to handle these variation tolerance concerns.

3.2 Strategy

To achieve our three design goals our circuit synthesis strategy relies on five key ideas. For clarity of exposition, in this section we enumerate these ideas as:

1. *Asymptotic Waveform Evaluation* (AWE) [Raghavan 93, Pillage 90] augmented with some simple, automatically generated analytical equations to predict circuit performance.
2. A non-linear constrained *optimization formulation* of the circuit synthesis problem that we map to an *unconstrained* optimization problem and solve via numerical optimization.
3. *Simulated annealing* [Kirkpatrick 83] as the optimization engine to provide robustness and the potential for global optimization in the face of many local minima.
4. A compiled database of industrial quality, non-linear device models, called *encapsulated device evaluators*, which provides the accuracy of a general purpose simulator while making the synthesis tool completely independent of low-level device modelling concerns.
5. A *relaxed D.C. formulation*, that avoids a CPU intensive D.C. operating point solution after each step in the optimization.

In the remainder of the chapter we discuss the motivation for these ideas and how they work together to reshape the equation-based synthesis strategy of Figure 3.1 and form a new, practical synthesis strategy.

3.2.1 Asymptotic Waveform Evaluation

Foremost, to streamline the path from a new topology to a sized circuit schematic, we do not require designers to derive, code, manipulate, or guide the symbolic pruning of the key circuit performance equations for the circuits they wish to synthesize. Instead, we replace the performance equations of Figure 3.1 with direct evaluation, *i.e.*, circuit simulation. Since SPICE-like general purpose simulation is too inefficient to be practical, we employ a recent innovation in simulation called Asymptotic Waveform Evaluation (AWE) [Raghavan 93, Pillage 90]. AWE is a robust, efficient approach to analysis of arbitrary linear RLC circuits that is several orders of magnitude faster than SPICE for small-signal analysis. By matching the initial boundary conditions and the first $2q-1$ moments of the actual RLC circuit transient response to a reduced q-pole model, AWE can predict circuit response using a reduced complexity model. Using the linearization techniques employed by circuit simulators (see

Section 3.2.4), we can linearize non-linear devices, generating RLC circuit models for the integrated circuits we wish to analyze. AWE can convert these circuit models into reduced-order transfer functions, which will yield the small-signal performance characteristics of interest to an analog designer. Thus AWE forms the core of a new performance prediction module that replaces the performance equations of Figure 3.1 at a fraction of the run-time cost of SPICE-like simulation, but is a general-purpose simulation technique that yields accurate results for any linearizable circuit.

3.2.2 Fully Automatic Synthesis Using an Optimization Formulation

To finish our re-design of the loop of Figure 3.1, we require a search mechanism to interactively perturb the evolving circuit design until it meets the performance specifications. Since we now explicitly avoid designer-supplied hints as to where a correct solution might lie, we cannot rely on search methods often used by equation-based tools—heuristic planning [Harjani 89, Degrauwe 89], rules [Trontelj 90, Klinke 92], qualitative reasoning [Toumazou 90], or other techniques that rely on circuit specific knowledge—to guide us to the solution. Thus, we return to solving the circuit design problem as an optimization problem, as originally suggested in [Rohrer 67] and discussed for simulation-based optimization in Section 2.3.

In Section 2.3, we formulated circuit design as a multi-objective non-linear constrained optimization problem. Because we will be optimizing with simulated annealing (see Section 3.2.3), which is an unconstrained optimization technique, we will

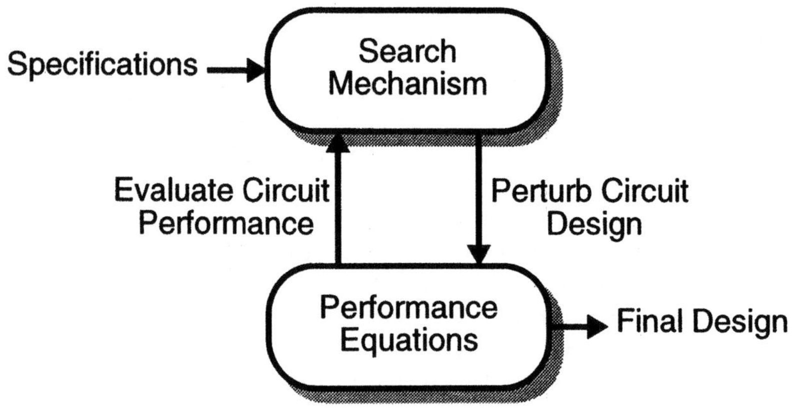

FIGURE 3.1 Search Process Used in Equation-Based Analog Synthesis Tools

need to modify the constrained optimization formulation of Equation 2.1 (repeated as Equation 3.1 for convenience):

$$\underset{\underline{x}}{\text{minimize}} \sum_{i=1}^{k} w_i \cdot f_i(\underline{x}) \qquad \text{s.t.} \quad \underline{g}(\underline{x}) \leq 0 \tag{3.1}$$

Recall that the set of objective functions $\underline{f}(\underline{x})$, is combined into a single function with the use of scalar weights, w_i. As an unconstrained technique, simulated annealing manipulates a single scalar function called a *cost function*, $C(\underline{x})$. To transform the constrained optimization problem into an unconstrained one, we can also apply weights to the constraint functions, $\underline{g}(\underline{x})$. This results in the following problem formulation:

$$\underset{\underline{x}}{\text{minimize}}\ C(\underline{x}) \tag{3.2}$$

$$C(\underline{x}) = \sum_{i=1}^{k} w_i f_i(\underline{x}) + \sum_{j=1}^{l} w_j g_j(\underline{x}) \tag{3.3}$$

where w_i and w_j are scalar weights as in Section 2.3.1. Note that the functions in $\underline{g}(\underline{x})$ are not in the same form as those in Equation 3.1. The constraints must be reformulated somewhat such that they contribute to the cost function *only* when the constraints are violated. However, there is a one-to-one correspondence between the constraints of Equation 3.1 and those of Equation 3.3, so for notational convenience we will refer to them simply as $\underline{g}(\underline{x})$.

Since it does not require any designer supplied hints to guide the search, this optimization formulation is well-matched to an AWE based performance evaluation core. Unfortunately, because there may be many local minima in the solution space of designable circuits [Temes 67, Nye 84, Harvey 92], we must address the issues of starting point independence and global optimization.

3.2.3 Simulated Annealing

Selecting the best optimization engine for our problem is not an easy task because of the difficulty of issues such as global optimization and starting point independence. In equation-based approaches it was possible to avoid some of these issues by using designer-supplied circuit-specific heuristics to guide the search. Since we have explicitly forbidden the hard-coding of any such circuit-specific heuristics in our synthesis tool, there is no choice but to use an optimization method that directly addresses global optimization and starting point independence. Although there are many techniques for local-optimization [Fletcher 90], the number of options for global optimization are more limited [Kan 89].

In selecting an optimization algorithm, we considered several important characteristics of several optimization algorithms. These characteristics include utility, ease of formulation, efficiency, starting point independence and ability to find a global optimum. After evaluating these techniques for our application, we selected simulated annealing as our optimization algorithm. We summarize our evaluation—concentrating particularly on simulated annealing—as follows:

- *Utility.* Since its introduction in 1983 [Kirkpatrick 83], simulated annealing has become very popular in the computer-aided design field, finding effective practical application in such difficult, diverse optimization problems as VLSI layout [Cohn 91, Basaran 93, Sechen 85], routing [Mitra 92], and others [Bonomi 84, Lam 86, Elgamal 84, Fleisher 85]. Many of these problems are NP Complete [Garey 79] and extremely difficult to solve using more traditional techniques. In contrast, other techniques such as genetic algorithms [Goldberg 89, Holland 92], response surface modeling [Guardiani 93], and homotopy [Richter 83, Trajkovic 91] are relatively unproven in the CAD field, although preliminary results are encouraging. Since the goal of our work is centered around design automation and not around global optimization *per se*, the proven utility of simulated annealing across a broad range of CAD problems is an important factor in our selection of simulated annealing for our application.
- *Flexibility.* Our optimization problem will be large and complex. Because we are performing synthesis, we anticipate having to minimize problems with tens of variables and hundreds of constraints. Many aspects of this problem will be difficult to control tightly: The device models are adopted from detailed circuit simulators, and the user will specify the circuit topology, performance metrics, and independent variables. For large, complex optimization problems of this kind, flexibility of formulation is essential. In this regard, the benefits of the simulated annealing algorithm include optimization without derivatives, ability to handle a flexible cost function, and robustness in the face of discontinuities. For our application, optimization without derivatives is very important. Obtaining derivatives will be difficult for several reasons: the complexity of the problem, the fact that many of the performance constraints will be unique to each synthesis problem and provided by the designer, and the inefficiency of numerical differentiation over a large number of variables. The ability to handle a flexible cost function is also important. Some optimization algorithms restrict the allowable mathematical form

of the cost function. This is unacceptable because the performance specifications for each new circuit problem are specified by the designer. Further, the encapsulated device evaluators will also affect the cost function and they are not written by us, but adopted from detailed circuit simulators. Finally, robustness in the face of discontinuities is also essential. Recall that simulation-based optimizers often failed because the simulator could not evaluate the performance of a circuit configuration. This failure led to a gap in the cost function, and gradient-based optimization techniques typically cannot succeed if the cost function is not connected. Simulated annealing does not suffer from these restrictions, can handle a wide range of cost functions, and does not require derivatives.

One potential drawback with using simulated annealing is that circuit design is most naturally formulated as a constrained optimization problem and we have formulated it as an unconstrained problem (see Equation 3.3). However, many optimization techniques also require unconstrained formulations. The problem is to ensure that $\underline{g}(\underline{x}) \leq 0$ when the optimizer terminates. To accomplish this, the weights w_j will have to be set carefully. Controlling the weights will indeed prove a substantial challenge, which we address in Section 4.5.2. However, it does not present any difficulties that are fundamentally intractable.

- *Efficiency.* In some circles, annealing is reputed to run slowly. This unfortunate reputation is probably due in part to comparisons between simulated annealing and *local* optimization techniques, which in general are more efficient. In contrast to this reputation, when compared to other *global* optimization techniques, simulated annealing often requires fewer calls to the function to be optimized. Recall that many of the applications that employ simulated annealing use it to solve NP complete problems that are extremely difficult to solve using more traditional techniques. Perhaps it is this last factor that has led to the perception that simulated annealing is a "last resort" to be used when the problem is otherwise intractable. Further, when compared with global optimization methods, such as homotopy [Richter 83] and even gridded random multi-start [Dixon 78, Kan 89, Horst 93, Schoen 93], simulated annealing appears in a fairly favorable light. Moreover, with the increasing research effort on the efficient implementation of annealing, for some problems it is now achieving run-times that compare favorably with tuned heuristic methods [Lam 88].
- *Starting Point Independence.* One key problem with early attempts to synthesize circuits using an optimization formulation was the inability to consistently determine an appropriate starting point [Rohrer 92]. Because simulated annealing generates moves in a pseudo-random fashion and incorporates controlled hill-climbing, it is essentially starting-point independent.

- *Global Optimization.* One important difficulty with generalizing an optimization method for circuits to a synthesis method for circuits is the problem of local minima. To address this problem, a global optimization method is needed. Although global optimization in general is not a solved problem, much progress has been made within the last 15 years [Kan 89] and several new techniques have emerged. Simulated annealing is one of these. It has a theoretically proven ability to find the global optimum under certain restrictions [Romeo 91].

  To provide a little insight into what this means, consider the fact that there are no finite algorithms for determining the true global optimum (say global minimum) of a function. This is easy to see. For a real valued objective function $f:\Re^n \rightarrow \Re$, the true global minimum, x_*, $x_* \in \Re^n$, can be defined as:

$$f(x_*) \leq f(x) \, \forall x \in \Psi \tag{3.4}$$

  Where the set $\Psi \subset \Re^n$, and Ψ is convex, compact and contains the global minimum as an interior point. Since the function is real valued, there are an infinite number of points that would need to be evaluated to show Equation 3.4 and this would require an infinite number of function evaluations. Thus, no finite algorithm can determine whether a point is truly the global minimum. In practice, a global optimization problem is usually formulated as the search for a member of a set within ε of the global minimum, for $\varepsilon > 0$. In the case of simulated annealing, proofs of the ability to find such a global optimum use this formulation and also stipulate certain constraints on the form of the cost function that typically cannot be enforced in practice. In particular, such proofs typically demonstrate convergence in probability σ as time approaches infinity for a suitably constructed annealing search algorithm, *i.e.*,

$$\lim_{t \rightarrow \infty} (\Pr\{C(x) - C(x_*) \leq \varepsilon\} \geq 1 - \sigma) \tag{3.5}$$

  Despite these restrictions, the proofs are not without meaning. Their existence shows an underlying theoretical power to the algorithm that has been verified by empirically good results for practical implementations and real problems.

The key advantages of simulated annealing—starting point independence and global optimization—outweigh the difficulties in its application to the circuit design problem. Thus, we select it is as the search mechanism to complete our re-design of the synthesis methodology in Figure 3.1.

3.2.4 A Library of Encapsulated Device Models

Although we have selected the two principal components of our new strategy — AWE and simulated annealing — there remain several important practical issues yet to be addressed. As with any equation-based or simulation-based analysis of a circuit, we use models to linearize any non-linear devices, generating a small-signal circuit that can be passed to AWE. Unfortunately, in a practical circuit synthesis system there is no longer any alternative to using industrial-strength device models. High-performance circuits rely on the device performance characteristics achievable in aggressive technologies. Simplified models of the current-voltage relationships at the terminals of these devices fail to capture many of these important characteristics. The real problem is not that industrial strength models are difficult to come by – indeed, they have long been recognized as a key aspect of detailed circuit simulation [Rohrer 92]. Unfortunately, supporting such powerful models is not easy within a synthesis environment. Not only does the complexity of these models make them difficult to work with in an equation-based environment, but also, existing models are written for simulation and map physical parameters to performance. For synthesis, performance is given and physical dimensions must be derived, rendering existing models unsuitable unless solved within an iterative loop.

A more subtle and far-reaching problem is the difficulty of determining D.C. bias points. Simulators use iterative methods such as Newton-Raphson [Vlach 94] to solve for the D.C. node voltages of a circuit. Instead of implementing this CPU-intensive portion of a general purpose simulator, most synthesis tools use simplified device models that allow them to analytically solve for the D.C. bias points of the circuit. This provides a substantial speed-up over simulation, and greatly reduces run-times but at the cost of prediction accuracy.

Coupled with the accuracy problems of hard-coding simplified non-linear device models into a synthesis tool is the fact that device modelling techniques continue to evolve. A single set of device models—even if they were not simplified—is not sufficient to appease every analog designer. In fact, device models are often closely guarded corporate secrets unavailable to synthesis tool authors outside the company.

To address these issues, our new strategy relies on a library of encapsulated device evaluators [Carley 93]. An encapsulated device evaluator treats each device as a black box. All aspects of the device's representation and performance are handled as requests to the evaluator. In this manner, the models are completely independent of the synthesis system and can be easily replaced or altered. For our purposes we rely entirely on device models adopted from detailed circuit simulators such as SPICE[Nagel 75].

3.2.5 Reformulating the Synthesis Problem

The use of encapsulated device evaluators does not solve the problem of determining the D.C. bias voltages in a circuit. We still require node voltages to feed to the device evaluators to determine performance characteristics. In fact, it is now necessary to use a numerical method to solve for them, since the internal details of the model are completely hidden. The most straightforward approach is to employ the same Newton-Raphson algorithms used in SPICE-like circuit simulators. However, since this iterative solution technique is a source of divergence problems in SPICE and consumes a substantial portion of the CPU time involved in detailed circuit simulation, it would be an important improvement if we can find an alternative approach. To this end, we re-cast the optimization formulation of our synthesis problem to our advantage. Recall our cost function, $C(\underline{x})$, given by Equation 3.3. The usual approach would be to solve the D.C. bias problem at each new circuit candidate $\underline{x}$ we visit during optimization and then evaluate $\underline{f}(\underline{x})$ and $\underline{g}(\underline{x})$. Instead we choose to explicitly formulate the Kirchhoff's law constraints implicitly solved during the D.C. biasing and include them in $\underline{g}(\underline{x})$, the other constraint functions in our optimization problem. Just as we must formulate optimization goals such as meeting gain or bandwidth constraints, we now formulate D.C.correctness as yet another goal to meet. We refer to this technique as the *relaxed D.C. formulation.*

Using this technique we can realize a substantial reduction in CPU time by amortizing the cost of the D.C. solution across the entire optimization process. However, it does lead to several difficult issues. First, we must be able to converge reliably to a realizable circuit. Recall that Maulik resolved this problem by using sequential quadratic programming techniques (SQP) to solve the constrained optimization problem. Because it is a constrained optimization technique, if the optimization converges, it converges to a D.C. correct solution. We take the formulation one step further, by using an unconstrained optimization method. We allow both D.C. goals and performance goals to evolve within a single cost function and balance their needs with scalar weights. As we shall see in Section 4.3, by incorporating a complete Newton-Raphson solver within the simulated annealing optimizer and letting the annealer decide when it is appropriate to force the circuit to be D.C. correct, we can achieve substantial speedup while attaining consistent convergence.

3.3 Architecture

We have developed five key strategies to address the shortcomings of previous approaches to analog synthesis and achieve our goals for a new circuit synthesis system. They are: AWE, synthesis through unconstrained optimization, simulated annealing, encapsulated device models, and the relaxed D.C. formulation. We now show how these strategies are combined to form an architecture for an analog circuit synthesis system. This architecture (see Figure 3.2) describes the path from an unsized analog circuit topology and a set of performance specifications to a completed, synthesized circuit. This path comprises three phases: *analysis*, *code generation*, and *solution*. We refer to the combination of the first two phases as *compilation*. The goal of compilation is to generate an executable performance predication module, customized for the particular circuit synthesis problem at hand. Code is generated that implements the cost function, $C(\underline{x})$, whose minimum value occurs at a circuit design that best meets the input specifications. The solution phase then uses this customized performance predictor and performs synthesis via optimization as discussed in Section 3.2.2.

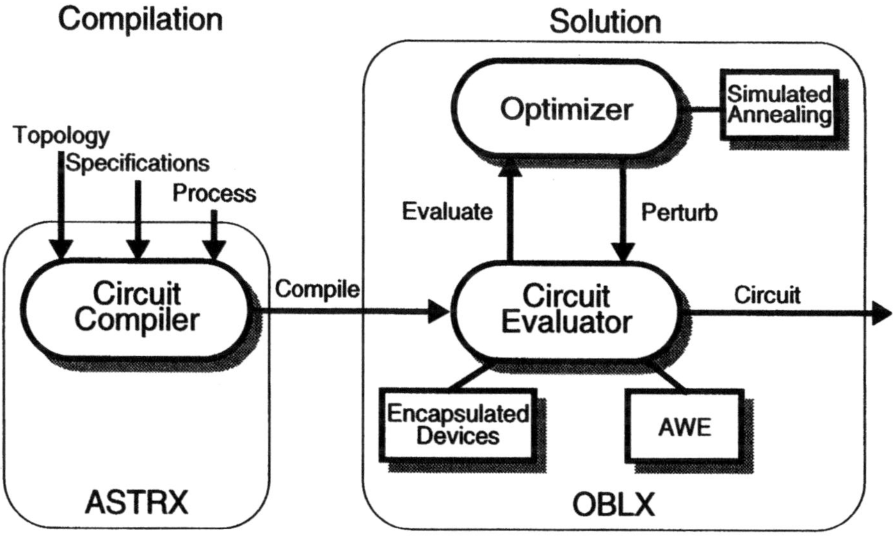

FIGURE 3.2 New Synthesis Architecture

3.3.1 Analysis and Code Generation

The synthesis process begins with a problem description created by the user. The first phase in our new synthesis architecture is analysis of this description. We automatically derive all the information needed by the performance prediction module to evaluate the simulated annealing cost function, the right hand side of Equation 3.3. We provide an overview of this process in this section and a full description in Chapter 5. The first step is to determine the set of independent variables, $\underline{x}$, most of which are simply given in the problem description. However, the compiler must also implement the relaxed D.C. formulation. This means that the D.C. constraints, *i.e.*, Kirchhoff's laws, are formulated explicitly and added to the cost function. This involves three compilation tasks. First, the compiler must analyze the circuit and determine a set of linearly independent nodes. Second, it must write analytic expressions for Kirchhoff's current law at each of these nodes and generate constraint functions from these expressions. And finally, the compiler must augment $\underline{x}$ with variables that will hold the voltages values for each linearly independent node. The annealer can then manipulate these node voltages as independent variables in order to satisfy Kirchhoff's laws. Completing Equation 3.3, we know that the objective functions, $\underline{f}(\underline{x})$ are included in the description. But, they must be translated into programming language form and, to evaluate performance characteristics of the evolving circuit to be evaluated, appropriate calls to the encapsulated device evaluators and AWE added. A similar process is necessary for the constraint functions, $\underline{g}(\underline{x})$. But, because of the relaxed D.C. formulation, $\underline{g}(\underline{x})$ also includes the Kirchhoff's current law constraints that were derived for each linearly independent node in the circuit.

3.3.2 Solution

During compilation, a performance prediction module is generated that can evaluate $C(\underline{x})$. This cost function has been carefully constructed so that its minimum value occurs when $\underline{x}$, the component and voltage values in the circuit, specifies an optimal design of the circuit to synthesize. This performance prediction module is compiled and linked to a "core" simulated annealing engine that numerically solves for the minimum of $C(\underline{x})$. As detailed in Chapter 4, by finding the minimum of $C(\underline{x})$, the solver determines the device sizes and D.C. bias point that best meets the description of the problem specified by the user.

The overall flow of this architecture is reminiscent of [Swings 91a], except that we automate the synthesis of the D.C. bias point (rather than ask the designer to add the relevant equations), use AWE to evaluate A.C. performance (rather than pruned symbolic equations), and use an efficient, aggressive annealing strategy (to evaluate 10^5 rather than 10^3 circuit candidates over a typical synthesis run).

3.4 A Design Scenario

So far, we have enumerated the ideas that underlie our new synthesis strategy and provided abstract motivation for their selection. We have also presented a high-level description of the path from circuit topology to sized schematic. In this section we present a complete synthesis example to make concrete the entire path from problem to solution. In our example, we trace the design process, attempting to provide insight into how the designer thinks about the problem, what input she needs to provide to the circuit compiler, what the compiler does to transform that problem into an optimization problem, and what the optimization engine must do to complete the design. We use this example to add focus to the synthesis ideas we have already presented and to provide a context for the more detailed discussions of the circuit compiler and optimization engine presented in Chapter 5 and Chapter 4 respectively.

3.4.1 The Circuit Design Problem

For clarity, we present a simple circuit design problem. Assume a designer wishes to design a differential amplifier like the one shown in Figure 3.3 for the specifications in Table 3.1. Her first task is to describe the circuit topology she wishes to have sized. Assume that she selects the circuit topology shown in Figure 3.4. In this topology, the resistor values are known to be 2kΩ, but the dimensions of the transistors and the value of the bias current are unknown. She can begin her problem description by making a list of these independent variables:

$$\underline{x} = \{W, L, I\} \quad . \tag{3.6}$$

Note that she wishes to determine only a single width and length because the circuit is fully differential, implying that the two transistors (M1, M2) must be matched. In general, there will not be a one-to-one correspondence between the variables and the

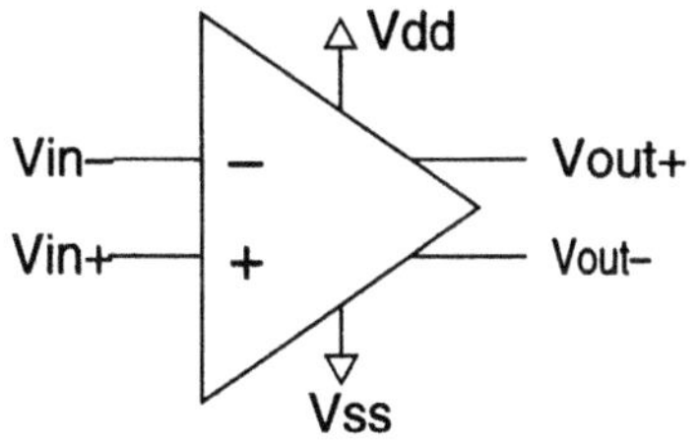

FIGURE 3.3 Design Example: A Differential Amplifier

devices in the circuit. Thus, she must also specify which variables apply to which devices. She can combine this with the topological description of the circuit by listing all the elements, their values, and their port nodes. For our example, this list is as follows:

TABLE 3.1 Design Example: Performance Specifications

Attribute	Specification
A_{dm}	maximize
UGF	≥ 1 MHz
A_{cm}	≤ 0 dB
SR	≥ 1 V/μs

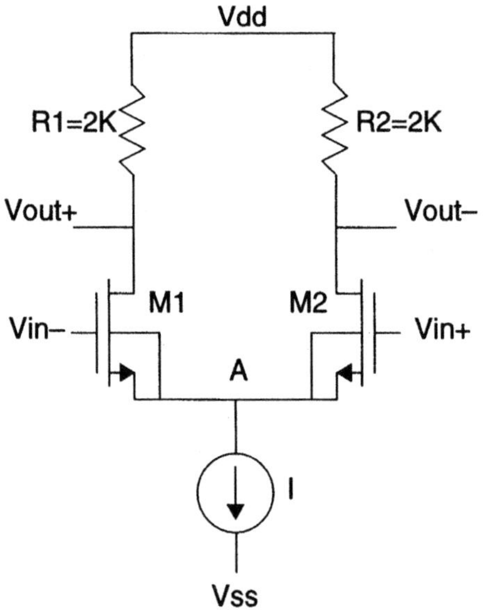

FIGURE 3.4 Design Example: Circuit Topology to Size

```
Amplifier circuit {
        R1      value=2kohms            nodes=(Vdd, Vout+)
        R2      value=2kohms            nodes=(Vdd, Vout+)
        M1      width=W, length=L       nodes=(Vout+,Vin–,A,A)
        M2      width=W, length=L       nodes=(Vout–,Vin+,A,A)
        I       value=I                 nodes=(A,Vss)
}
```

Note that specifying matching devices involves simply reusing the same variable both for the widths of both devices and for the lengths of the devices.

After describing the circuit topology and variables that she wishes the tool to determine, the designer's next task is to describe the performance specifications. From Table 3.1, the specifications of interest are: A_{dm}, the differential gain of the amplifier; UGF, the unity gain frequency; A_{cm}, the common mode gain; and SR, the slew rate. Although these are fairly commonplace specifications for an amplifier (and only a small subset of the specifications an industrial design would require), it is instructive to define them further. We can define differential gain as

$$A_{dm} = \frac{(\text{Vin+}) - (\text{Vin–})}{(\text{Vout+}) - (\text{Vout–})}, \tag{3.7}$$

unity gain frequency as the frequency at which

$$|A_{dm}| = 1, \tag{3.8}$$

common mode gain as

$$A_{cm} = \frac{(\text{Vin+}) + (\text{Vin–})}{(\text{Vout+}) + (\text{Vout–})}, \tag{3.9}$$

and slew rate as the maximum value of

$$\frac{d}{dt}\text{Vout+}(t)\,. \tag{3.10}$$

Given a simple circuit, symbolic expressions can be derived for these expressions. However, for more complex amplifier designs, these values must be found through simulation[1]. To do this, a designer simulates a circuit that puts the amplifier into a test configuration that makes the specification easy to measure. To measure A_{dm}, UGF, and A_{cm}, we can use the test circuits shown in Figure 3.5. In each case, the gain is measured as simply (Vout/Vin). Measuring UGF with the circuit on the left requires increasing the frequency of Vin until (Vout/Vin) is 1. More importantly, we may need more than one circuit for a set of specifications, and, in order for the measurements to be meaningful, we must load the amplifier in a fashion the simulates the other circuits that the completed design is intended to drive (in this case with Cl). Thus, the designer supplies an environment, or *test jig* [Burr 90], in which the test can be made.

Since our synthesis system uses AWE to simulate circuit performance, the designer must provide a similar *simulation oriented* definition of each specification. To describe fully a specification, we need to provide a test jig, a set of measurements to make on the jig, and any simple arithmetic that must be performed to calculate the values we are interested in. Similar information is required to simulate the performance of the circuit in any simulator, so giving the specifications to the synthesis tool in this fashion is in fact quite convenient. For example, we can summarize the specification for A_{dm} as follows:

```
Adm test jig {
        Use an amplifier circuit
        Cl+     value=1pF           nodes(Vout+, gnd)
        Cl–     value=1pF           nodes(Vout–, gnd)
        Vin     value=1V A.C.       nodes(Vin+, Vin–)
        calculate Adm as Vout/Vin
}
```

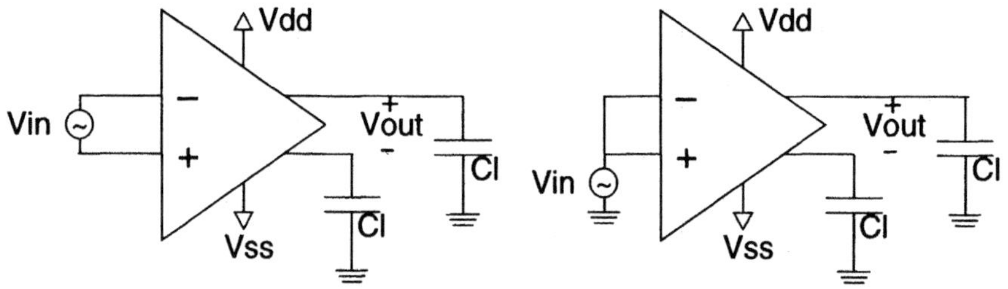

FIGURE 3.5 Simulation Test Jigs for Measuring Vdm (left) and Vcm (right)

1. Or by building the circuit and measuring. Regardless, this would still involve test jigs, which is the point we are trying to make.

AWE actually determines transfer functions such as Vout/Vin directly, so it is ideally suited for the kind of gain calculation shown in line 6.

For our example, we can reuse the same jig for UGF, but will need an additional test jig for A_{cm}. Until now, we have neglected the slew rate. This is because measuring slew rate would require a transient simulation, which is not possible with AWE[1]. However, unlike gain and unity gain frequency, slew rate is readily described with an easily derived expression. If we assume that we are interested only in the rate at which the output slews downwards, we can write this expression by inspection as:

$$\mathrm{SR} = \frac{I}{2\,(\mathrm{Cl} + \mathrm{Cd})} \tag{3.11}$$

where Cd is the capacitance at the drain of one of the transistors. Thus, the designer needs to be able to include both equation-based and simulation-based descriptions of specifications in the description of the synthesis problem.

The designer's description of the synthesis problem is almost complete. However, she needs to include one additional circuit, a bias circuit, before our circuit compiler can transform the problem into a form suitable for our optimization engine. To understand the need for this circuit, we must discuss the treatment of non-linear devices in more detail. Like the algorithms at the core of most circuit simulators, AWE can predict circuit performance only for linear circuits. As a result, the synthesis system must provide a mechanism to linearize the transistors in the designer's circuit, replacing them with a linear model that can then be processed by AWE. This task is performed by the encapsulated device evaluators. For a MOS transistor and a bipolar junction transistor (BJT), this process is shown in Figure 3.6. The evaluator converts the dimensions and node voltages of the device into a set of element values for the complex linear circuit model shown on the right. After replacing each transistor with its model, we can then use AWE to evaluate the circuit's performance. The problem then is to make sure we know all the inputs to the evaluators for each transistor.

Because the width and length are part of our independent variables, they are readily available, although we will need to re-evaluate them for each sizing solution we visit. However, the evaluators also require the voltages at the ports of the transistors. If we attempt to solve for these voltages by solving for the node voltages in one of the test jigs of Figure 3.5, we may run into a problem. Because these circuits are in high gain, open loop configurations, we may run into numerical problems when we attempt to

1. For more details about the potential for transient simulation with AWE see [Alaybeyi 90].

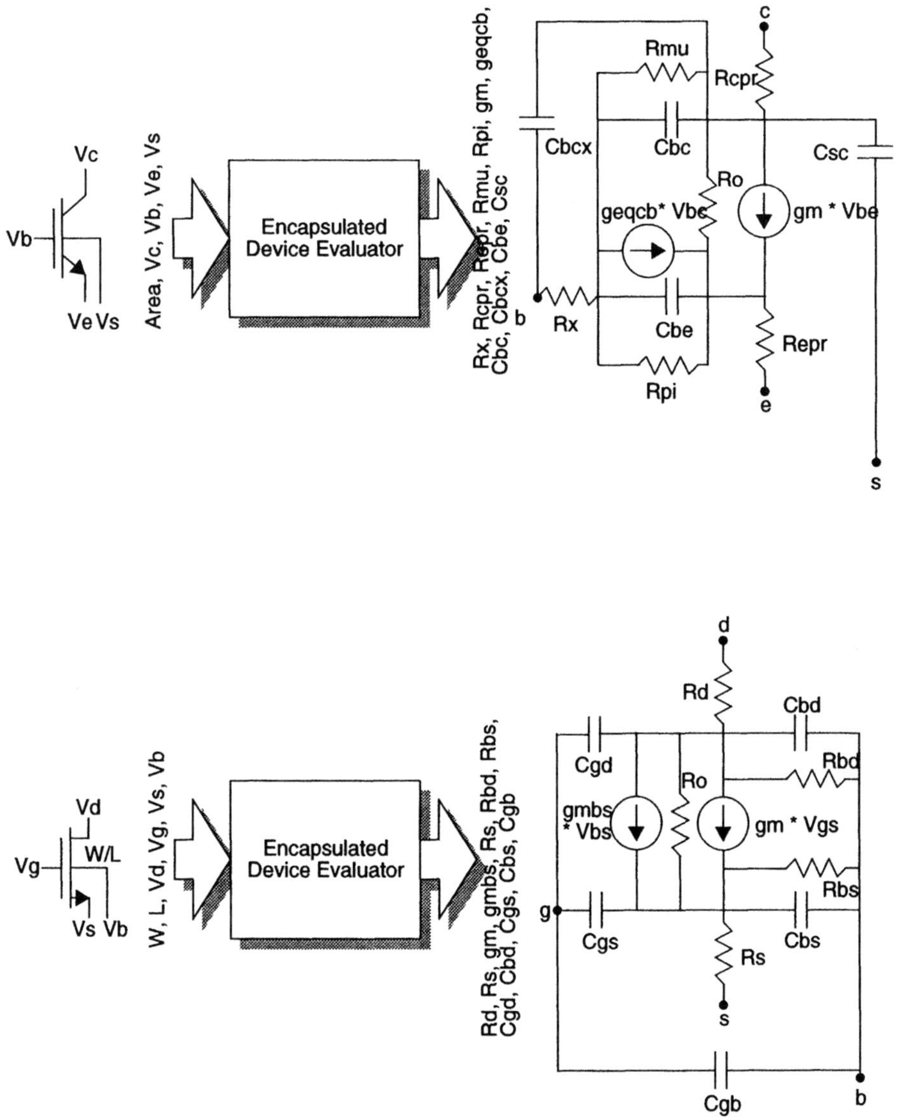

FIGURE 3.6 Two Encapsulated Device Evaluators and their small-signal models. (top) BJT. (bottom) MOSFET.

solve for their node voltages. (For the moment, we defer the more fundamental problem of being able to solve for the node voltages in a circuit with non-linear devices.) These test jigs have been designed to measure directly small-signal performance characteristics. This is appropriate because we are using AWE to evaluate performance, and AWE handles linear circuits or circuits that have been linearized at a single operating point, *i.e.*, non-linear circuits that have been converted to small-signal circuits. Thus, we have two choices: alter our test jigs so that they can be used for both biasing and AWE or use a separate circuit for biasing. For detailed circuit simulators, performance evaluation and biasing are always performed with a single test jig. The advantage is that this most closely duplicates the process of wiring up a physical test circuit. However, physically unrealistic simulation "tricks" are often used to allow the circuit to bias properly without adversely affecting the performance measurements. For example, the circuit in Figure 3.7 is often used to determine A_{dm}, where Cbig and Rbig are usually picked to be impracticably large. Unfortunately, this circuit introduces a low frequency zero into the system, and this can complicate performance evaluation with AWE. As a result, we have designed our synthesis system such that the node voltages for non-linear devices are taken from a *separate and distinct* bias circuit. For our example, we can use the bias circuit of Figure 3.8. This circuit is selected because it biases the amplifier at a typical D.C. operating point, so that the voltages we pass to the encapsulated device evaluators are those that we would find in a quiescent amplifier before a small-signal input is applied. The topology of this bias circuit is the last piece of input the designer must supply to the circuit compiler.

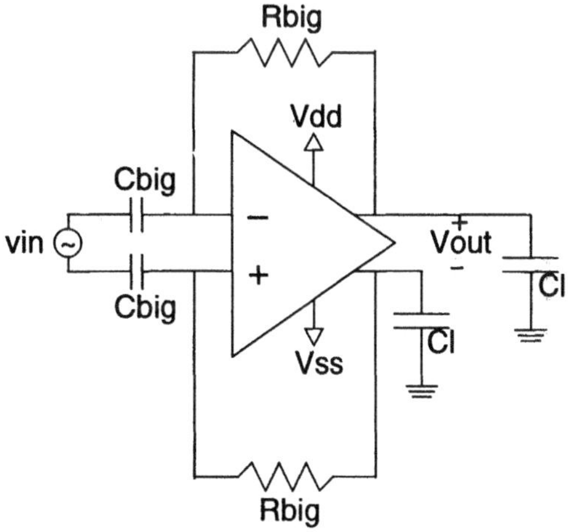

FIGURE 3.7 A_{dm} Test Jig that Biases Properly

3.4.2 Mapping the Synthesis Problem Into An Optimization Problem

The next step in the synthesis path is performed by the circuit compiler. It must transform the problem description into a cost function whose minimum is the optimal solution to the circuit design problem. Recall that the basic form for the cost function is given by Equation 3.3,

$$C(\underline{x}) = \sum_{i=1}^{k} w_i f_i(\underline{x}) + \sum_{j=1}^{l} w_j g_j(\underline{x}) , \qquad (3.12)$$

where $\underline{x}$ is the set of independent variables, $f_i(\underline{x})$ is one of a set of objective functions, $g_j(\underline{x})$ is one of a set of constraint functions, and w_i and w_j are scalar weights. In this section, we map the problem description of our example into the components of this cost function.

We begin with the independent variables. This is quite simple: we use the independent variables specified by the designer. Thus,

$$\underline{x} = \{W, L, I\} . \qquad (3.13)$$

However, we shall see in a moment that we will need to add other variables to $\underline{x}$ as well.

The next step is to determine $\underline{f}(\underline{x})$, the set of objective functions. For our example, the only objective is to maximize A_{dm}. Thus, $\underline{f}(\underline{x})$ contains only the function that

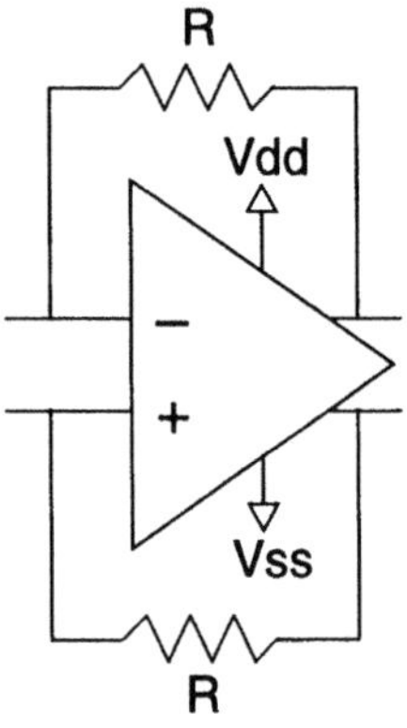

FIGURE 3.8 Design Example: Bias Circuit

calculates A_{dm}, which we label $f_{Adm}(\underline{x})$. (We defer until Section 4.4.1 the minor detail that we wish to *maximize* $f_{Adm}(\underline{x})$ but we wish to *minimize* $C(\underline{x})$.)

We are now ready to describe the additions we need to make to $\underline{x}$. To do so, we trace the information required to calculate $f_{Adm}(\underline{x})$. Recall that A_{dm} is actually Vout/Vin, and we use AWE to calculate this transfer function directly from the test jig in Figure 3.5. However, AWE requires a linearized circuit, so we need to replace the transistors in that test jig with small-signal models. As shown in Figure 3.6, we can use the encapsulated device evaluators to determine the values for the linearized-models, but we need the device dimensions and port voltages as input to the evaluators. The dimensions are already part of $\underline{x}$, so they are readily available, but we still need the port voltages. Our solution is to simply include the voltages of the nodes in the circuit in $\underline{x}$. Thus, for our example, we need to add independent variables for the linearly independent set of node voltages in the bias circuit of Figure 3.8. Now, $\underline{x}$ is:

$$\underline{x} = \{W, L, I, \text{Vout+}, \text{Vout–}, V_A\}. \tag{3.14}$$

Note that some of the nodes in the circuit have fixed voltage values, so these node voltages are not included.

This is the key step towards the relaxed D.C. formulation. In one sense this formulation greatly simplifies the problem because the node voltages are then readily available to be used by the device evaluators. However, we must now complete the relaxed D.C. formulation by finding some means to force the node voltages to take values such that Kirchhoff's laws are obeyed. To accomplish this for a non-linear biasing circuit, we must again turn to the encapsulated device evaluators. Similar to the process we used to generate small-signal models, we can use the evaluators to generate large-signal models like the one in Figure 3.9. If we replace the transistors in the circuit we are trying to design (Figure 3.4) with the large-signal models returned by the device evaluators, we end up with the circuit of Figure 3.4. Note that the values of the current sources we have just added are not fixed, but are determined by the device evaluators and may change when the device dimensions or port voltages change. What we have gained by using these large-signal models is the ability to write equations that enforce Kirchhoff's laws. Kirchhoff's current law (KCL) can be written at each node in the circuit in a quite mechanical fashion. If we substitute the amplifier large-signal

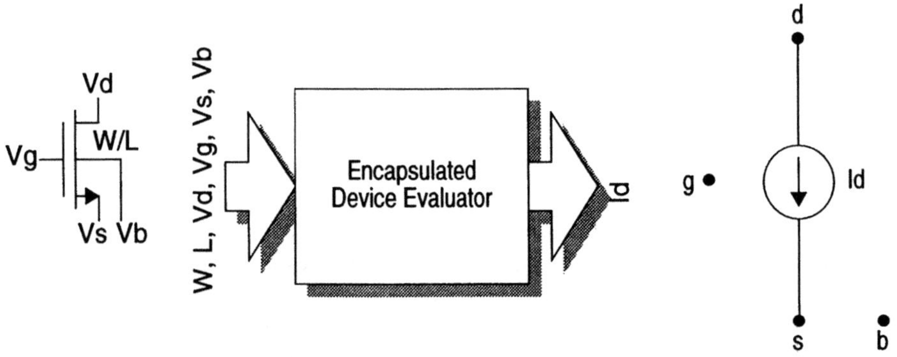

FIGURE 3.9 Encapsulated Device Evaluators, Simplified D.C. Large-Signal Model of a MOSFET

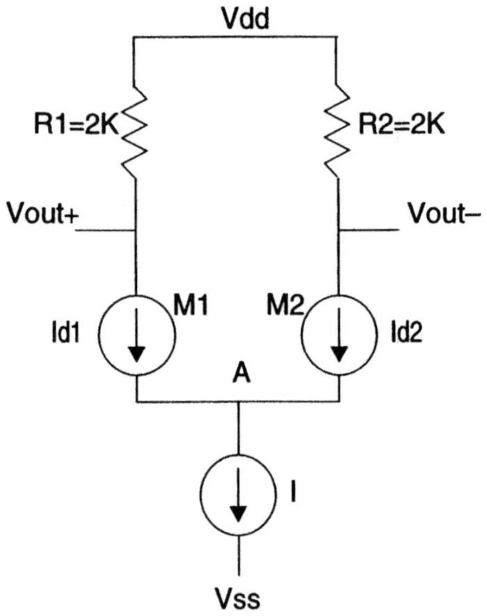

FIGURE 3.10 Design Example: Large Signal Model of the Amplifier

model of Figure 3.4 into the bias circuit of Figure 3.8 and sum currents at the nodes, we obtain the following equations:

$$\text{at node A: } \mathrm{I} - \mathrm{Id1} - \mathrm{Id2} = 0 \tag{3.15}$$

$$\text{at node Vout+: } \frac{(\mathrm{Vout+}) - \mathrm{Vdd}}{2000} + \mathrm{Id1} = 0 \tag{3.16}$$

$$\text{at node Vout-: } \frac{(\mathrm{Vout-}) - \mathrm{Vdd}}{2000} + \mathrm{Id2} = 0 \tag{3.17}$$

Because these constraint equations act to enforce KCL, they must be met when our optimization is complete—if we have a valid circuit. To make this happen, we include the KCL equations in the set of constraint equations, $\underline{g}(\underline{x})$, in our cost function, $C(\underline{x})$. Specifically, we can re-write Equation 3.15 as the constraint equation:

$$g_{\mathrm{A}}(\underline{x}) = \max(0, |\mathrm{I} - \mathrm{Id1} - \mathrm{Id2}| - \tau_{abs}) \tag{3.18}$$

Thus, $g_{\mathrm{A}}(\underline{x})$ contributes a penalty to the cost function whenever the KCL error at node A is larger than some numeric tolerance, τ_{abs}[1]. We formulate the other KCL equations in the same fashion, creating $g_{\mathrm{Vout+}}(\underline{x})$ and $g_{\mathrm{Vout-}}(\underline{x})$. Together, these three constraints form the other half of the relaxed D.C. formulation.

So far, we have used $\mathrm{A_{dm}}$ to define the only member of $\underline{f}(\underline{x})$ and used the KCL equations derived for the relaxed D.C. formulation to define three members of $\underline{g}(\underline{x})$. For our example, the other members of $\underline{g}(\underline{x})$ correspond to the other three performance specifications we wish to design for. Thus, we need constraint functions for UGF, $\mathrm{A_{cm}}$, and SR. Including these final three terms, we obtain the final form of the cost function we need to optimize to complete the circuit design:

$$\begin{aligned} C(\underline{x}) = {} & w_{Adm} f_{Adm}(\underline{x}) + w_{Acm} g_{Acm}(\underline{x}) + w_{UGF} g_{UGF}(\underline{x}) + \\ & w_{SR} g_{SR}(\underline{x}) + w_{\mathrm{A}} g_{\mathrm{A}}(\underline{x}) + w_{\mathrm{Vout+}} g_{\mathrm{Vout+}}(\underline{x}) + w_{\mathrm{Vout-}} g_{\mathrm{Vout-}}(\underline{x}) \end{aligned} \tag{3.19}$$

1. This formulation of the KCL constraint is somewhat simplistic, but we defer further details to Section 4.4.5

3.4.3 Solving the Optimization Problem

In this section we take a brief look at the last step in the path from unsized circuit topology to complete design: optimization. We optimize the cost function of Equation 3.19 by repeatedly changing one or more of the variables in $\underline{x}$ and re-evaluating $C(\underline{x})$. We defer a detailed discussion of the weights in the cost function and of the simulated annealing algorithm we use for optimization until Chapter 4. Instead, we focus on the steps that must be performed each time the cost function of Equation 3.19 is evaluated. Starting with the device dimensions, linear element values, and node voltages in $\underline{x}$, we perform the following steps:

1. Use the device evaluators to generate linearized small-signal versions of the two AWE test jig circuits (in Figure 3.5).
2. Use AWE on the test jigs to regenerate the transfer functions for A_{dm}, A_{cm}, and UGF.
3. Compute A_{dm}, A_{cm}, and UGF from transfer functions. Compute SR by substituting component values from the device evaluators in the expression provided by the designer. Insert the new specification values in the cost function.
4. Use the device evaluators to generate a large-signal model version of the bias circuit (Figure 3.8).
5. Use the KCL equations derived for the bias circuit to calculate current error at each of the three nodes where we do not know the voltage. Insert these errors to complete the final three terms of the cost function.

3.5 Revisiting Unresolved Issues

We have presented a fairly detailed example of the process by which our new synthesis strategy determines circuit component values for a circuit topology and performance specifications. However, there are issues that we have yet to address. Now that we have seen a simplified example, there are many details of the formulation and algorithms used that we have not adequately described. We defer discussion of these to Chapters 4 and 5. The example also highlights several conceptual issues related to the relaxed D.C. formulation, these include evaluating the performance of circuits that do not satisfy Kirchhoff's laws and guaranteeing convergence. We discuss these more conceptual issues in this section.

The first issue to address is what it means to evaluate a circuit that is not D.C. correct. Assuming convergence, at the end of the synthesis process, the circuit will be D.C.

correct (within tolerances not unlike those used in circuit simulation), but early in the process the sum of the currents entering a given node may be substantial. This is a non-physical situation and difficult to conceptualize. One way to view this circuit is to imagine an additional current source at each node. This current source sources or sinks the current required to ensure D.C. correctness. The goal of the Kirchhoff's law constraints added to $\underline{g}(\underline{x})$ is to reduce the value of these currents sources to zero.

An alternative view is to recall that the node voltages are inputs to the encapsulated device evaluators and attempt to quantify the effect errors in these voltages have on circuit performance prediction. Recall that an independent set of node voltages is contained in $\underline{x}$. For the purposes of our discussion, we separate $\underline{x}$ into two subsets, $\underline{x}_v$, the node voltage variables, and $\underline{x}_d$ the variables that hold the other device parameters. We also assume (without loss of generality) that the circuit contains only devices modelled by encapsulated evaluators. Thus, $\underline{x} = \{\underline{x}_v, \underline{x}_d\}$. Then, let the modelling performed by the set of encapsulated device evaluators be $\underline{\beta}(\underline{x}_v, \underline{x}_d)$. and the performance of a given circuit be $\wp(\underline{\beta}(\underline{x}_v, \underline{x}_d))$. If we are not D.C. correct, $\underline{x}_v$ will have some error $\Delta\underline{x}_v$, and the error in our measure of performance will be:

$$\Delta\wp = \wp(\underline{\beta}(\underline{x}_v + \Delta\underline{x}_v, \underline{x}_d)) - \wp(\underline{\beta}(\underline{x}_v, \underline{x}_d)) . \quad \textbf{(3.20)}$$

Thus, we can consider the presence of D.C. error as simply an error factor in our performance modelling. This error factor allows us to visit many more possible circuit configurations within a given period of time, albeit modelling the performance of each a little less accurately. As the optimization proceeds, $\Delta\underline{x}_v \to 0$, $\Delta\wp \to 0$, and the performance prediction becomes as accurate as if complete solves were done at each step.

With this goal in mind, the choice of annealing is also somewhat apt. Annealing's hill-climbing abilities have encouraged experimentation with a variety of problem formulations with the same flavor as relaxed D.C. correctness, *i.e.*, particular terms of the cost function to be minimized correspond to penalties that quantify the distance from the current solution to a feasible one. For example, in 2-D integrated circuit layout, a placement of cells on the IC is *in*feasible if any design rules are violated, *e.g.*, if any of the cells overlap. A standard formulation of this problem for simulated annealing [Jepsen 84] allows intermediate placements to violate these design rules, but penalizes this violation in the cost function. This is a proven technique used in many circuit layout tools [Cohn 91, Sechen 87]

It is worth considering a seeming inconsistency in our overall synthesis formulation. We clearly cannot guarantee that each annealing move visits a physical (KCL correct) circuit. Nevertheless, we evaluate this probably nonphysical circuit using detailed device models and highly accurate AWE techniques. It might well be argued that this accuracy is wasted, and clearly it is not fully exploited early in the optimization. However, there is considerable information in these evolving circuits—even when $\Delta \underline{x}_v$ is large—that might be lost if simple device models evolved into more detailed models as more accuracy is required. Practically, this evolution would be difficult because different models often predict significantly different performance and changing models might substantially alter the performance predictions. In fact, much of the information gained from these "non-physical" circuits is very useful because the annealer can still learn much from them. For example, if we need to achieve more gain in our circuit, we probably need to increase the g_m on some critical device. We do not need a precise D.C. bias point to know that we must either increase that device's width, its current, or both. We can successfully make coarse decisions such as this even without a precise D.C. bias point. During the annealing process, as the magnitudes of these current sources are forced to zero, the interactions between devices are more precisely modelled and our decisions become more accurate.

The second issue we must address is convergence. Clearly, if the Kirchhoff's current law (KCL) constraints are being violated when the optimization completes, the circuit's performance will not match the synthesis tool's predictions. It would be reassuring if we could offer some mathematical guarantee that we can always converge to minima where the KCL constraints are met. However, this is not the case. Unfortunately, there is little that we can show rigorously, and we must rely on empirical measures of convergence reliability to gauge the effectiveness of our formulation. We argue that this should not be viewed as a substantial weakness of our method. Many numerical techniques, like Newton-Raphson or simulated annealing, are effective on real design problems but have poor or difficult to prove theoretical properties. Nevertheless, circuit simulators that employ Newton-Raphson and device layout tools that employ simulated annealing have proven to be effective CAD tools. Thus, the lack of a formal convergence guarantee should not be viewed as a major drawback of our new synthesis architecture.

3.6 Summary

In this chapter we have followed the evolution of a new analog circuit synthesis architecture. This architecture is based on the critical analysis of previous analog synthesis formulations and the willingness to trade an increase in run-time for substantially lessened preparation by a human designer. The architecture relies on five key ideas: AWE as the core of the performance predictor, fully automatic synthesis using an optimization formulation, simulated annealing as the optimization engine, a library of encapsulated device evaluators and a relaxed D.C. formulation. We then presented an overview of the architecture that shows how these five strategic ideas can be combined into a cohesive system architecture, and walked through a comprehensive example that showed the path through the system. Finally, we concluded the chapter considering the theoretical implication of the relaxed D.C. formulation and the lack of a theoretical proof of convergence. The next two chapters describe in detail the remainder of the conceptual, engineering, and implementation decisions that have been made in order to prove the viability of this architecture. For ease of explanation, we reverse the order of the synthesis flow and present the solution phase before compilation.

CHAPTER 4

Synthesis Via Annealing

In this chapter we present the algorithms used to solve the synthesis problem. In general, we use simulated annealing to optimize a complex, continuous-valued cost function over $\Re^n$ whose minimum corresponds to the optimal circuit design, given the topology and performance specifications provided by the user.

4.1 Overview of Simulated Annealing

To solve our circuit synthesis problem, we minimize a cost function with simulated annealing. Before we describe the simulated annealing algorithm in detail, we motivate its basic structure. Originally introduced by Kirkpatrick [Kirkpatrick 83], simulated annealing derives its name from a useful analogy that can be drawn between the annealing of solids and the optimization of properties of large, complex systems. Annealing is a controlled cooling process that involves melting a substance, then carefully cooling it until a highly regular crystalline structure is formed. Making an analogy to the optimization of large complex systems, variables in a system correspond to atoms in the substance being annealed and a minima in the optimization process corresponds to the highly regular crystalline structure that results from annealing. The key innovation in simulated annealing is the introduction of the concept of an *effective temperature*, T, into optimization. As with annealing, the simulated annealing process begins with a high effective temperature and slowly reduces it. When the effective temperature has decreased such that the system being optimized has settled into a minima, the optimization process is complete.

Essentially, SA is optimization via random search. The key difference is that new states are accepted not only when they cause the cost to improve, but, with a probability that is a function of temperature, when they cause the cost to worsen. This allows a progression from a locally good state through a worse state to the global minimum; a capability known as *hill-climbing*. Given an initial problem state, $\underline{x}_0$, and an initial temperature, T_0, a pseudo-code description the SA algorithm is shown in Figure 4.1.

At the core of the algorithm are the two functions *generate* and *accept*. How the annealer generates a new problem state given the current state is encoded in the function *generate* (on line 4 of Figure 4.1). The *accept* function on line 5 determines whether the newly generated problem state replaces the current problem state. For simulated annealing, the *accept* function is called the *Metropolis criterion* [Metropolis 53], and is defined in Equation 4.1.

$$random() < e^{\left[\frac{C(\underline{x}) - C(\underline{x} + \Delta\underline{x})}{T}\right]} \quad \textbf{(4.1)}$$

Here, *random* is a function that returns a uniform random number on the interval [0,1]. The behavior of the metropolis criteria is shown conceptually in Figure 4.2. Here, *downhill* moves are moves to a new state with a lower cost than the present state, while *uphill* moves increase the cost. For a downhill move, $C(\underline{x}) - C(\underline{x} + \Delta\underline{x})$ is positive, and the exponent in Equation 4.1 will be positive, so the function on the right of the inequality will yield a value larger than one. Since the range of *random* is [0,1], these downhill moves are trivially accepted. The probability of accepting an

```
1  x = x0 , T = T0
2  while not frozen(x, T)
3      while not done_at_temperature(x, T)
4          x + Δx = generate(x)
5          if accept(x, x + Δx, T) then
6              x = x + Δx
7      T = update_temp(T)
```

FIGURE 4.1 The Simulated Annealing Algorithm

uphill move depends on the ratio of the change in the cost function to the present temperature. For very *hot* temperatures, this ratio is small and almost all uphill moves are accepted. When the temperature has cooled such that it is only *warm* (on the left in Figure 4.2), the probability of accepting a large uphill move is substantially reduced. When the temperature is reduced further, the probability of accepting uphill moves also decreases (on the right in Figure 4.2). This probability continues to drop until almost no uphill moves are accepted. At this point, the problem settles into a minima, the bottom of a valley in the cost surface, and is considered to be *frozen*.

The functions *frozen, done_at_temperature* and *update_temp* determine when the annealing is finished, how many moves should be performed at each temperature and how the temperature is updated, respectively. The initial temperature, T_0, and these critical control functions are collectively referred to as the *cooling schedule*. Optimal cooling schedules have been the subject of much research [Lam 88a, Aarts 87, Huang 86, Strenski 91] and selection of an appropriate schedule is crucial to an efficient implementation.

In the remainder of this chapter, we present a complete description of our annealing formulation by expanding on the algorithm of Figure 4.1. This description details the four principal components of the algorithm:

- $\underline{x}$, the problem representation, which maps the circuit being designed to the current state within the annealer;
- *generate*, which implements the annealer's *move-set*, the set of possible perturbations the annealer can use to manipulate the current state;

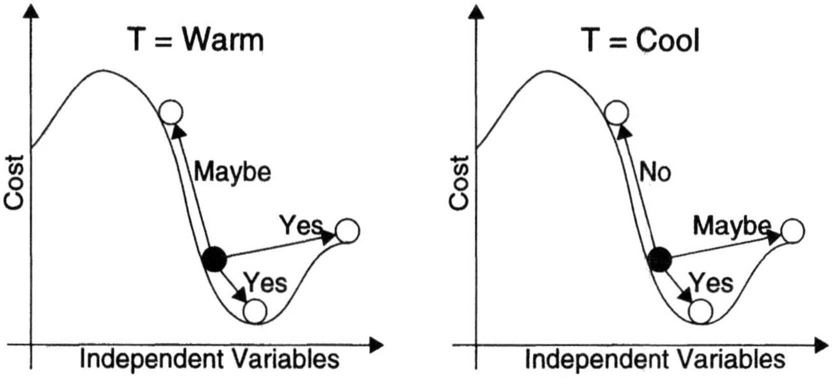

FIGURE 4.2 Effect of Temperature on Likelihood of Move Acceptance With Metropolis Criteria. (left) When warm. (right) When cool.

- $C(\underline{x})$, the cost function, which determines how the cost is calculated and what its components are;
- and the *cooling schedule,* which controls T, directing the overall cooling process.

In the sections that follow, we describe each of these components in sequence.

4.2 Problem Representation

Recall from Chapter 3 that we represent the circuit synthesis problem as an optimization problem, and that the goal of this optimization problem is to minimize the cost function $C(\underline{x})$ given by Equation 4.2.

$$C(\underline{x}) = \sum_{i=1}^{k} w_i f_i(\underline{x}) + \sum_{j=1}^{l} w_j g_j(\underline{x}) \tag{4.2}$$

Each of the parts of this function corresponds to specific aspects of the synthesis problem we are trying to solve, and finding the global minimum of $C(\underline{x})$ corresponds to finding the optimal sizing/biasing of the given topology to meet the user-specified specifications. To describe our problem representation we define the following parts of $C(\underline{x})$:

- Let $\underline{x}$ be the set of n independent annealing variables. Part of this set of variables we refer to as *user-specified.* These are the physical geometries of semiconductor devices or component values of passive circuit components that the user wishes the synthesis tool to determine. The other part of this set we refer to as *node voltage variables.* These are the values of node voltages on a linearly independent set of nodes in each of the bias circuit topologies in the synthesis problem. As discussed in Section 3.4, they are required as part of the relaxed D.C. formulation. For each variable, x_i, a minimum value, x_i^{min}, and a maximum value, x_i^{max}, is specified by the user, *i.e.*, $x_i \in [x_i^{min}, x_i^{max}]$.
- Let $f_i(\underline{x})$ be the ith member in the set of k objective functions, $\underline{f}(\underline{x})$. Each objective function corresponds to a user-specified circuit performance objective. The details of this correspondence are presented in Section 4.4.
- Let $g_j(\underline{x})$ be the jth member in the set of l constraint functions, $\underline{g}(\underline{x})$. The constraint functions are derived from the circuit synthesis problem in several ways. The most important of these are performance constraints and D.C. constraints. The performance constraints are similar to performance objectives, except that, unlike

objectives, once a constraint is met, further improvement of that performance metric does not affect the cost function. D.C. constraints implement the relaxed D.C. formulation. As discussed in Section 3.4, each D.C. constraint quantifies the Kirchhoff current law error at a particular node. Again, the details of the constraint functions are presented in Section 4.4.

In addition to the components of $C(\underline{x})$, there are several aspects of the internal representation of the circuit synthesis problem that we will refer to throughout this chapter. For convenience we collect these definitions as:

- Let Y be a nodal admittance matrix for a linear circuit. This matrix defines the current/voltage relationship:

$$Y\underline{v} = \underline{i} \tag{4.3}$$

  A Y matrix can be generated from the set of linear elements in a circuit by a process called *stenciling*. Stenciling is a straightforward algorithm that uses an element type and its node connections to determine the locations in the matrix where the element's value must be added or subtracted. The annealer generates and maintains a separate Y matrix for each bias circuit in the synthesis problem description.
- Let A be a circuit matrix AWE requires to perform circuit performance analysis. Like Y matrices, an A matrix can be generated from the set of linear elements by stenciling. The annealer generates and maintains a separate A matrix for each AWE test jig circuit in the synthesis problem.
- Let $\underline{v}$ be the set of node voltages in a bias circuit.
- Let $\Delta\underline{i}$ be the set of Kirchhoff-current-law-violating currents at all the nodes in a bias circuit. Thus, the element in the set for a given node is computed by summing the currents at that node.

In the rest of this section, we concentrate on the set of annealing variables: how they are represented within the annealer and how they correspond to component values in the circuit under design.

Recall that some of the annealing variables, $\underline{x}$, are specified by the user in the problem description, and the remainder are derived during *compilation* as described in Section 5.2.2. As shown in Figure 4.3, the user may specify variables as one of two types: *logarithmic* and *continuous*. The user must also specify several attributes of each variable she defines. For each variable, x_i, a minimum value, x_i^{min}, and a maximum value, x_i^{max}, is specified by the user, and the annealer enforces

$x_i \in [x_i^{min}, x_i^{max}]$ by bounds checking the variables throughout the annealing process. Typically the bounds given should be very relaxed to allow the annealer as much freedom as possible in determining variable values.

Logarithmic variables are used when it makes sense to select the variable's value from a discrete set. For example, device sizes can reasonably be regarded as discrete quantities. The underlying limitations are either lithographic (how accurately we can etch a device) or reflect the need to restrict designs to a known set of carefully qualified device sizes for which accurate models are available. Moreover, there is considerable advantage to be had from properly discretizing device sizes: the coarser the discretization, the smaller the space of reachable sizes that must be explored. Because small changes in device sizes make proportionally less difference on larger devices, in our formulation we typically use logarithmic variables for device dimensions. For logarithmic variables, the annealer maintains the variable value as an integer. As shown in Figure 4.3, this integer then acts as a look-up index in a table of real values. A separate table is generated for each logarithmic variable, using the user-specified variable bounds and a user-selected grid coarseness. The grid coarseness determines the number of values per decade stored in the look-up table.

Unlike device sizes, node voltage values must clearly be continuous to determine an accurate bias point. Highly nonlinear devices can be extremely sensitive to very small changes in their operating point. For highly accurate performance prediction, we are aware of no voltage discretization scheme that is viable for analog applications. The choice to implement the problem representation within the annealer as a mix of discrete and continuous variables is somewhat unusual and will make other aspects of

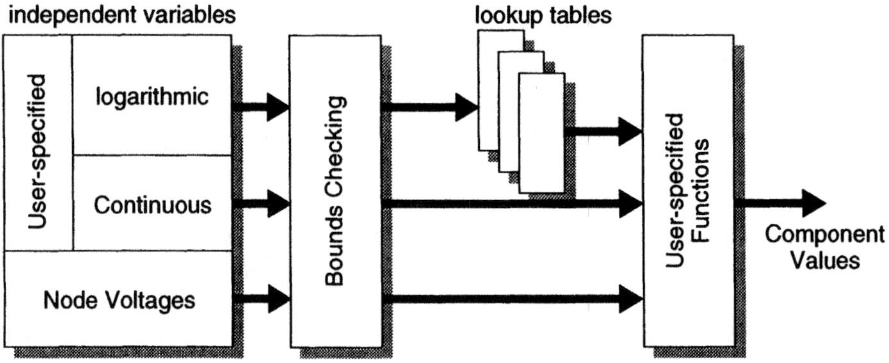

FIGURE 4.3 Mapping from Independent Variables ($\underline{x}$) to Circuit Component Values

the annealer design somewhat unusual as well (for example, the *generate* function described in Section 4.3).

The last step in mapping the internal representation of the circuit to the actual component values is a set of user-specified functions. In practice, these functions are typically identity functions that map a single variable in $\underline{x}$ directly to one or more MOSFET widths, MOSFET lengths, node voltages, resistor values, etc. For ease of discussion, unless otherwise noted, we shall assume that each user-specified variable maps directly to one or more passive circuit element values or active device parameters. Thus, we shall use the terms component value and variable value interchangeably.

4.3 Move Generation

The second component of the annealer design is the *generate* function. Given the present state, $\underline{x}$ we must determine the *move-set*, the set of allowable perturbations on $\underline{x}$. The basic problem is the need for a mechanism to generate each perturbation $\Delta\underline{x}$ that is both coarse enough to explore a large portion of the cost surface early in the annealing process and fine enough to allow the annealer to find a precise final solution. For discretized variables this is straightforward. There is always a smallest allowable move, an *atomic* perturbation, and the only decision is how to add larger moves to the move-set for efficiency. For example, for a device dimension, we could rely only on moves that altered the variable by 0.01μ, although this would be remarkably inefficient. Ideally, we need a mechanism that dynamically adds larger moves when they are appropriate for the current state of the annealing problem.

However, because we include continuous variables in our problem representation, the move-set design is even more complex. Simulated annealing was originally designed for combinatorial optimization, where all the variables are discrete and the solution space countable. For an *n*-dimensional real-valued state, $\underline{x} \in R^n$, it is difficult to determine the correct *smallest* $\Delta\underline{x}$ because adjustments in real values may be infinitesimally small. Near the end of the annealing process, when D.C. constraints are no longer very relaxed, we may need to explore voltage changes with the accuracy of a detailed circuit simulator. But it is unlikely we can afford the time required to search for a feasible bias point if we restrict ourselves solely to moves of 1 nV in magnitude, the accuracy used in a typical simulator. Hence, for continuous variables we need to determine how small the smallest move should be and again to automatically adjust the mix between large and small moves.

We are aware of several attempts to generalize SA to problems with real-valued variables [Gelfand 91, Bilbro 91, Vanderbilt 84, Szu 87] that present methods of controlling the move-set. Two are of particular note: The *topography-following* approach of Vanderbilt and Louie [Vanderbilt 84] focuses on the second moments of the random walk formed by a sequence of accepted annealing moves generated on an n-dimensional cost surface. Vanderbilt and Louie show that the covariance matrix generated by these moves can be used to gauge the shape of the "valley" currently being explored. This shape can in turn be used to control the range and minimum size of moves for the real-valued variables. In contrast, the *tree-annealing* approach of Bilbro [Bilbro 91] implicitly builds a model of the n-dimensional cost surface in the form of an n-dimensional tree. The distance from root to leaf is deeper in the regions where more successful solutions have been found. The tree is incrementally refined after each successful move, and is used to bias the generation of random moves toward more promising regions of the space. Although both methods show promise, there is a lack of data for problems with large dimensionality, *i.e.*, more than 10 or 20 variables. Moreover, the matrix manipulations required for the Vanderbilt and Louie topography-following scheme may computationally dominate the overall annealing run-time, a situation we would prefer to avoid. There is also concern about the storage requirements and numerical properties of these matrix manipulations. Storage is also a fundamental problem with Bilbro's approach. Here, the tree stores all the problem states generated during the annealing process and thus grows inexorably with each accepted move.

An alternative to these two methods is to use problem specific information to aid in selecting the correct largest and smallest moves. For analog circuit synthesis, this is actually quite straightforward. For user-specified variables, such as the size of a resistor or capacitor, the precision with which the value can be set is a function of the underlying fabrication process and typically readily available to the designer. For node voltages, the smallest moves are dictated by the desired accuracy of performance prediction. Duplicating the method used in detailed circuit simulators, these tolerances can be specified with an absolute and relative value and the smallest allowable move derived from them. Thus, we can use problem specific information to determine the largest and smallest allowable moves for each variable.

We must then decide when to use large moves and when to use small ones. Clearly, larger moves are more appropriate during the early stages of the optimization when it is important to explore as much of the cost surface as possible. Smaller moves become essential when exploring the valley where the final solution is located so it can be determined exactly. Using this intuitive explanation, other implementations of annealing [Swartz 90] have actively controlled the move size, contracting the allowable range exponentially as a function of decreasing temperature. However, the rate at which the move ranges are contracted is somewhat arbitrary and typically requires input from the user. Further, some researchers have found that large moves are occasionally very important *late* in the annealing process [Cohn 92]. In an attempt to design a more flexible move control mechanism that uses the statistical behavior of the problem as the optimization proceeds to control the move generation process, we have designed an automatic move control mechanism based on work with a discrete annealing problem by Hustin [Hustin 87]. This mechanism requires an enumeration of all allowable *move classes*. The mechanism then biases the move selection process by assigning probabilities to each move class. We discuss the details of this process in the next few paragraphs.

To begin, we define a set of *move classes* for each variable. Each class is a tuple (x_i, r) where x_i is a variable to be perturbed by this move class and r is a range $[r^{min}, r^{max}]$ related to $[x_i^{min}, x_i^{max}]$ and the accuracy tolerance, ε_i, of that variable. For example, for a voltage variable with range $[0, 10]$ and ε_i of 1×10^{-6}, we might create three move classes with ranges $[-5\mu V, 5\mu V]$, $[-5mV, 5mV]$, and $[-5V, 5V]$. The number of move classes can also be controlled by the user, although it is not crucial and a small number like three is a good default. The idea is that during annealing, we will randomly select a move class. This determines not just x_i, the variable to perturb, but r the range with which to bound the perturbation. Once selected, we generate Δx_i as a uniform random number in $[r^{min}, r^{max}]$. Finally, Δx_i may be adjusted to ensure the new variable value, $x_i + \Delta x_i$, lies within $[x_i^{min}, x_i^{max}]$.

In addition to these purely random, *undirected* move classes, we augment the annealer's move-set with *directed* move classes that follow the D.C. gradient. This technique is similar to the theoretically-convergent continuous annealing strategy used by Gelfand and Mitter [Gelfand 91]. Gelfand and Mitter use gradient-based moves within an annealing optimization framework. They formulate moves as a step in the direction of the gradient, except that a random "noise" factor is added to both the direction and size of the step. As the temperature cools, successively smaller amounts of noise are used and the optimization becomes purely gradient descent. In [Gelfand 91], they then prove that this technique converges to a global minima given certain restrictions on the problem. Our addition of D.C. gradient moves has the same flavor. We incorporate gradient-based voltage steps as part of our set of move classes. Although we have no theoretical proof of convergence, we find these substantially improve the efficiency of the annealer at finding node voltages that satisfy Kirchhoff's laws. Using KCL-violating currents calculated at each node, $\Delta \underline{i}$, and the factored small-signal matrix, Y^{-1}, we can calculate the voltage steps as:

$$\Delta \underline{v} = \alpha (Y^{-1} \Delta \underline{i}) \tag{4.4}$$

where α is a scaling factor that bounds the range of the move. Thus, this move affects all the node voltage variables simultaneously. Equation 4.4 also forms the core of a Newton-Raphson iterative D.C. solution algorithm, which is incorporated as an atomic move in the annealer's move-set.

The complete algorithm for our implementation of Newton-Raphson is shown in Figure 4.4. The algorithm is fairly lengthy, and the following points are worth discussing further. First, the algorithm begins with the independent variables ($\underline{x}$), and the circuit node voltages ($\underline{v}$). Although $\underline{v}$ can be derived directly from $\underline{x}$, we distinguish them in the algorithm for clarity. Second, starting on line 4, we use ζ^k to denote the set of branches incident on node *k*, and I_b is the current in branch *b*. Also, the symbols τ_{rel}, τ_{abs}, κ_{rel}, and κ_{abs} refer to constants that control the numerical tolerance on the convergence of the algorithm. Similar constants are also used in detailed circuit simulators.

In summary, the complete move-set comprises classes of these types:

- **Undirected moves** where we modify a single variable and generate Δx_i as a uniform random number in $[r^{min}, r^{max}]$.
- **Directed moves** where the Newton-Raphson algorithm is used to perturb all the node voltage variables simultaneously. We distinguish several move classes of this form by limiting the size and number of steps the algorithm may take. As a result, some Newton-Raphson moves search for the nearby D.C. correct solution more tenaciously than others.
- **Combination moves** that perturb a user-specified variable in an undirected fashion and then immediately perform a complete or partial Newton-Raphson solve. Moves of this type are designed to be able to alter the circuit and circuit performance without adversely affecting the amount of D.C. bias error in it. Because of the relaxed D.C. formulation, when the optimization is nearly frozen it is difficult for the annealer to adjust circuit component values without incurring a high cost as a result of changes in the D.C. operating point. These combined moves combat that problem.

Because there may be a large number of variables, the move-set for a particular circuit may be very complex. For a typical circuit, each variable will have several undi-

Start with variables $\underline{x}$ and node voltages $\underline{v}$.

devs ← Update_Encapsulated_Devices($\underline{x}$, $\underline{v}$)

foreach node, k

$$\Delta i_k = \sum_{b \in \zeta^k} I_b$$

$$mag_k = \frac{1}{\|\zeta^k\|} \cdot \sum_{b \in \zeta^k} |I_b|$$

$$kclerr_k = |\Delta i_k| - (\tau_{rel} \cdot mag_k + \tau_{abs})$$

$$dverr_k = 0$$

while $(kclerr > 0) \vee (dverr > 0)$ and not over iteration limit

  Y ← Stencil_Linearized(*devs*)

  Y ← Stencil_Linear($\underline{x}$)

  $\Delta \underline{v} = Y^{-1} \Delta \underline{i}$

  $saved = \underline{v}$

  until $\sum_{k \in nodes} \Delta i_k$ decreases or we reach the attempt limit

    $\underline{v} = saved + \alpha(\Delta \underline{v})$

    decrease α

    devs ← Update__Encapsulated_Devices($\underline{x}$, $\underline{v}$)

    foreach node, $k, \Delta i_k = \sum_{b \in \zeta^k} I_b$

  if $\sum_{k \in nodes} \Delta i_k$ did not decrease, return failure

  $kclerr_k = |\Delta i_k| - (\tau_{rel} \cdot mag_k + \tau_{abs})$

  $dverr_k = |\Delta v_k| - (\kappa_{rel} \cdot v_k + \kappa_{abs})$

if $(kclerr \le 0) \wedge (dverr \le 0)$, return success

FIGURE 4.4 Newton-Raphson Algorithm

rected move classes and several combination move classes. Thus, the annealer must select from a palette that may contain several hundred move classes.

To bias the selection of move classes that are most appropriate to the particular state of the annealing process, we divide the overall annealing run into statistical intervals. Each statistical interval begins with a small number of move attempts of each class. During this collection phase, we track how each move has contributed to the overall cost function, and compute a *quality factor* that quantifies the performance of that move class. For move class i, the quality factor Q_i is given by Equation 4.5.

$$Q_i = (1/\|G_i\|) \sum_{j \in A_i} |\Delta C_j| \qquad (4.5)$$

Here, $\|G_i\|$ is the number of generated move attempts that used class i over the previous temperature and A_i is the set of accepted moves that used class i (a subset of the generated moves, *i.e.*, $A_i \subseteq G_i$). Further, $|\Delta C_j|$ is the absolute value of the change in the cost function due to move attempt j. Q_i will be large when the moves of this class are accepted frequently ($\|A_i\|$ is large), and/or if they change the value of the overall cost function appreciably (some $|\Delta C_j|$ are large).

Based on quality factors, we can compute the probability that a particular move class should be selected. If Q is the sum over all the quality factors, $Q = \sum_i Q_i$,

$$p_i = \frac{Q_i}{Q} \qquad (4.6)$$

can then be regarded as the fraction of moves that should be dedicated to move class i. In practice, we divide the annealing run into S *statistical intervals*, where each statistical interval is simply a number of contiguous moves. At the beginning of each statistical interval, we generate a small number of moves of each class to ensure that we continue to collect up-to-date statistics for each move class regardless of the probability assigned to it by Equation 4.6. Further, to ensure that the probabilities evolve in a controlled fashion, probabilities for the remaining portion of a statistical interval are set based on quality factors collected over a small number of previous intervals. Using this scheme, we automatically bias towards the moves that are most effective during this phase of the annealing run. Initially, when almost any move is accepted, large moves will change the cost function the most, giving them the largest quality factors and likelihood of being selected. Near the end of the annealing process, more small moves will be accepted and their quality factors and probabilities will increase.

Now that we have completed out description of the problem representation and of the generate function, we can turn our attention to the cost function.

4.4 Cost Function

The heart of the annealer is the circuit specific cost function, $C(\underline{x})$, implemented by the performance prediction module, which is the third of the four components of the annealing algorithm and maps each visited circuit configuration $\underline{x}$ to a scalar cost. Our cost function has the form:

$$C(\underline{x}) = \underbrace{C^{obj}(\underline{x})}_{objective} + \underbrace{C^{con}(\underline{x}) + C^{awe}(\underline{x}) + C^{dev}(\underline{x}) + C^{dc}(\underline{x})}_{penalty\ terms} \tag{4.7}$$

Where each term in Equation 4.7 represents a group of related terms whose composition is discussed below. Observe that there are two distinct kinds of terms: the *objective* terms are to be minimized, while the *penalty* terms are to be driven to zero. These correspond directly to the form of $C(\underline{x})$ in Equation 4.2, *i.e.*, $\underline{f}(\underline{x})$ has become the objective term and $\underline{g}(\underline{x})$ the penalty terms.

In the next five subsections, we describe each of the groups of terms in Equation 4.7: the objective terms (C^{obj}), the constraint terms (C^{con}), the AWE terms (C^{awe}), the device terms (C^{dev}), and the relaxed D.C. terms (C^{dc}). We then examine three other aspects of the cost function calculation: the algorithmic flow used to calculate the cost function, how AWE is used to calculate circuit transfer functions, and how AWE transfer functions are used to calculate circuit performance parameters.

4.4.1 Objective Terms

We begin our discussion of the terms in the cost function with a look at the objective terms. Following [Nye 88], circuit performance and figures of merit such as power and area can be specified as falling into one of two categories: an *objective* or a *constraint*. Regardless of the category, the user is also expected to provide a *good value*, and a *bad value* for each specification. These are used both to set constraint boundaries and to normalize the specification's range. Formally, let

$$\Omega^{obj} = \{f_i(x) \mid 1 \le i \le k\} \tag{4.8}$$

be the set of k performance objective functions provided by the user. Then Ω^{obj} is transformed into C^{obj} as follows:

$$C^{obj}(\underline{x}) = \frac{1}{\|\Omega^{obj}\|} \cdot \sum_{i \in \Omega^{obj}} \hat{f}_i(\underline{x}) \quad (4.9)$$

where,

$$\hat{f}_i(\underline{x}) = \frac{f_i(\underline{x}) - good_i}{bad_i - good_i} \quad (4.10)$$

and $f_i(\underline{x})$ is the ith specification function provided by the user, and bad_i and $good_i$ are the *bad* and *good* normalization values specified by the user for function i. These good and bad values provide an intuitive way for the designer to specify the normalization factors needed by the optimizer.

The user has a wide range of predefined functions at her disposal to create $f_i(\underline{x})$. These include node voltages, device model parameters, and functions to derive linear performance characteristics such as gain and bandwidth from the transfer functions derived by AWE for each circuit considered by the optimizer. (A complete description of these capabilities is given in Section 5.1, while a description of how AWE works and how it is used to calculated performance characteristics is presented in Section 4.4.7.) For example, to minimize (active device) area, the user might write the specification function:

$$f(\underline{x}) = \text{dev\_area}() \quad (4.11)$$

and give *good* and *bad* values of 10000μ^2 and 25000μ^2 respectively. If a particular circuit configuration had active devices whose area totaled 15000μ^2, this would contribute $(15000 - 10000)/(25000 - 10000)$, or about 0.33, to the overall cost function. This normalization process has several advantages. First it provides a natural way for the designer to set the relative importance of competing specifications. Second, it provides a straightforward way to normalize the range of values that must be balanced in the cost function. Note that the range of interesting values between *good* and *bad* maps to the range $[0, 1]$, and regardless of whether the goal is maximizing or minimizing $f(\underline{x})$, optimizing towards *good* will always correspond to minimizing the normalized function $\hat{f}(\underline{x})$. Finally, this normalization and the inclusion of the normalizing factor $(1/\|\Omega^{obj}\|)$ helps keep the cost function formulation robust over different problems, by averaging out the effect of having a large number of objectives for one circuit and a small number for another.

4.4.2 Constraint Terms

The treatment of constraints is quite similar to that of objectives. Formally, Let

$$\Omega^{con} = \{g_i(x) \mid 1 \le i \le l\} \tag{4.12}$$

be the set of l performance constraint functions provided by the user. Ω^{con} is mapped to C^{con} as:

$$C^{con}(\underline{x}) = \frac{1}{\|\Omega^{con}\|} \cdot \sum_{i \in \Omega^{con}} w_i \cdot \max(\hat{g}_i(\underline{x}), 0) \tag{4.13}$$

Where $\hat{g}_i(\underline{x})$ makes use of user supplied functions and is defined as

$$\hat{g}_i(\underline{x}) = \frac{g_i(\underline{x}) - good_i}{bad_i - good_i} \tag{4.14}$$

This is analogous to the definition of $\hat{f}(\underline{x})$ in Equation 4.10. However, constraints differ from objectives in two respects. First, negative values of $\hat{g}_i(\underline{x})$ are not included in the sum in Equation 4.13. This implies that although it is important for a constraint to achieve its *good* value, evaluating how much better than good is unimportant. If a constraint is met, it has no effect on the cost function. In contrast, the annealer is always trying to improve the value of an objective, regardless of how good it gets. The second difference is the inclusion of a weight factor w_i that biases the contribution of each constraint to the overall cost function. These weights are determined algorithmically during the annealing process and should not need to be adjusted by the user. We defer detailed discussion of this algorithm until Section 4.5.2. Weight values are typically larger than 1, which reflects the notion that it is more important for performance constraints to achieve their *good* values than it is for objectives. In fact, a typical strategy for setting *good* values for objectives is to make them unattainable, while *good* values for constraints should be within reach.

4.4.3 Numerical Robustness Terms

C^{awe} addresses some practical aspects of computing circuit performance with AWE. In particular, because of the relaxed D.C. formulation and random nature of the annealing exploration process, the linearized circuits passed to AWE for evaluation may be non-physical. As a result, AWE occasionally does not return a viable transfer function or cannot compute a transfer function at all. When this happens, the performance prediction module cannot derive all the information required to complete the cost function. If the difficulty is a low-order transfer function, C^{awe} will include a penalty that reflects the fact that the other performance characteristics are estimates based on insufficient information. If AWE cannot compute a transfer function at all, the annealer simply rejects the current move ($\underline{x} + \Delta\underline{x}$) and generates a new proposed move to continue. This ability to handle gaps in the cost surface is another major advantage of using simulated annealing. C^{awe} also includes small penalty terms that guide the problem state into regions that are more numerically stable. For example, the encapsulated device models may return small-signal element values on the order of 1×10^{-50}, while other elements in the circuit are 50 orders of magnitude larger. This extremely wide dynamic range can cause the admittance matrix to become numerically ill-conditioned. Small penalty terms in C^{awe} are triggered when the small-signal element values generated by an encapsulated device evaluator seem unreasonable. In practice, these terms and C^{awe} in general quickly dies to 0 and helps only to guide the annealer in the very early stages of the optimization search when it is searching for a feasible region from a completely random starting point.

4.4.4 Region of Operation Terms

The penalty terms C^{dev} are simple one-sided constraint functions specified by the user and are meant to force non-linear devices into a particular region of operation. For example, we typically constrain MOS devices to operate in strong inversion by writing:

$$V_{gs} - V_T \geq 0.2\text{V} \tag{4.15}$$

In part this conservative constraint reflects the lack of confidence in the weak-inversion region of any of our device models. Further, for linear analog circuits, we often force MOS devices to operate in the saturation region. But for switches, this is clearly inappropriate. Inclusion of these constraints is not essential, but we find they aid both in convergence and in manufacturability of the resulting design. By providing another indicator that the annealer is exploring the wrong part of the design space, the device penalty terms help to steer it to a reasonable solution. By constraining the feasible region to force MOS devices to be strongly in the saturation region, the resulting designs gain a measure of robustness in the face of mild variations in process parameters. Of course, the more complete treatment of variations in the second part of this book provides a more elegant and more complete solution.

When C^{dev} constraints are written for an active device, they must be rearranged to allow the constraint errors to be quantified. To do this, the actual constraint equation is converted into a relation of the form in Equation 4.16:

$$g_i(\underline{x}) \leq 0 \tag{4.16}$$

This conversion is done symbolically during the compilation process by collecting terms on the right-hand side of the equation and possibly multiplying by -1 to flip the comparison to less-than. As with the constraint terms, the final value of C^{dev} is computed from the set of device constraints, Ω^{dev}, where

$$\Omega^{dev} = \{g_i(x) \mid 1 \leq i \leq m\} \tag{4.17}$$

is the set of m device constraints that have been derived from the equations provided by the user. Then, C^{dev} is computed with Equation 4.13.

$$C^{dev}(\underline{x}) = \frac{1}{\|\Omega^{dev}\|} \cdot \sum_{i \in \Omega^{dev}} w_i \cdot \max(g_i(\underline{x}), 0) \tag{4.18}$$

Because the constraints are put into the form of Equation 4.16, negative values of $g_i(\underline{x})$ indicate that the constraint is being met. Also, as with performance constraints, scalar weights are used to balance competing device constraints.

4.4.5 Operating Point Terms

The last of the five components of the cost function—and arguably the most unusual—is the C^{dc} penalty term. This term implements the relaxed D.C. formulation of Section 3.2.5. Recall that we explicitly formulate Kirchhoff's laws as constraints—specifically, we formulate Kirchhoff's current law at each linearly independent node in all the bias circuits. These constraints form the C^{dc} term for the circuit we are synthesizing. Because they are constraints, we do not require Kirchhoff's laws to be satisfied after every move, but amortize the run-time cost of satisfying these laws over the entire optimization. At the end of the optimization, the C^{dc} terms will be zero and Kirchhoff's laws will be met.

C^{dc} includes two views of Kirchhoff's current law (KCL), a current view, C^{KCL} and a voltage view, C^{DV}. To calculate these terms, we first make the following definitions. Given a node k and its set of incident branches ζ^k, we define:

$$\Delta i_k = \sum_{b \in \zeta^k} I_b \tag{4.19}$$

$$mag_k = \frac{1}{\|\zeta^k\|} \cdot \sum_{b \in \zeta^k} |I_b| \tag{4.20}$$

Equation 4.19 computes the sum of the currents leaving the node (KCL). However, we need to make additional computations to define numerical tolerances for the KCL calculations. The first of these additional calculations (Equation 4.20) determines the magnitude of currents flowing through the branches at this node. The KCL term for node k, C_k^{KCL}, can then be calculated as follows:

$$\begin{aligned} err_k &= |\Delta i_k| - (\tau_{rel} \cdot mag_k + \tau_{abs}) \\ C_k^{KCL} &= \max(0, err_k) \end{aligned} \tag{4.21}$$

Essentially, the C_k^{KCL} term at node k is zero if the sum of currents incident on node k is less than τ_{abs} or less than $\tau_{rel} \cdot mag_k$. The tolerances τ_{abs} and τ_{rel} ensure that numerical truncation when adding the magnitude of a small current to the magnitude of a large one does not adversely affect the measure of KCL correctness. τ_{abs} is set to the desired absolute current accuracy, typically on the order of 10^{-9} amps, while τ_{rel} is the desired relative accuracy, typically on the order of 10^{-3}. This mechanism has a similar flavor to the stopping criteria used for the D.C. solution algorithm in detailed circuit simulators such as ASTAP [Astap 73].

The C^{DV} terms present an alternative view of KCL. They measure the voltage change required at each node to make the circuit satisfy KCL. By this definition, calculating the DV terms in C^{dc} exactly would require actually solving for the correct circuit node voltage values, which is what we are trying to avoid with the relaxed D.C. formulation. Instead, we estimate the DV error terms by multiplying the factored small-signal nodal admittance matrix, Y^{-1} by the error currents:

$$\Delta \underline{v} = Y^{-1} \Delta \underline{i} \tag{4.22}$$

Since the circuit will typically contain non-linear devices, the Y matrix will be a linearized estimate of the admittance taken at the present bias point. Thus, the DV errors will be first order approximations to the actual voltage error at each node in the circuit, and this estimate will improve as the circuit approaches D.C. correctness. As the annealing state approaches freezing, the DV terms will be quite accurate, but during the early stages of the anneal, the KCL terms will be the only reliable estimates of D.C. correctness because they are not approximations. The DV terms are somewhat more complicated to calculate than the KCL terms, and the complete algorithm, which is actually a subset of the Newton-Raphson algorithm (Figure 4.4), is shown in Figure 4.5. Because of the extra time taken for the calculation, and because KCL and DV are simply different views of KCL, the C^{dc} terms originally did not include DV terms explicitly. However, we found that for high impedance nodes the KCL tolerances had to be extremely tight to guarantee D.C. correctness, while more relaxed tolerances sufficed for low impedance nodes. The addition of the DV terms takes the impedance into account, providing a different perspective on the D.C. correctness of the circuit. Further, designers tend to have more intuition regarding voltage errors than current errors, so DV errors provide more useful feedback to designers.

Finally, given the set of KCL terms, Ω^{kcl} and a set of DV terms Ω^{dv}, the overall cost C^{dc} is calculated using Equation 4.13.

$$C^{dc}(\underline{x}) = \frac{1}{\|\Omega^{kcl}\|} \cdot \sum_{i \in \Omega^{kcl}} C_i^{KCL} + \frac{1}{\|\Omega^{kvl}\|} \cdot \sum_{j \in \Omega^{kvl}} C_j^{KVL} \quad \textbf{(4.23)}$$

devs ← Update__Encapsulated_Devices($\underline{x}$, $\underline{v}$)
foreach node, k

$$\Delta i_k = \sum_{b \in \zeta^k} I_b$$

Y ← Stencil_Linearized(*devs*)
Y ← Stencil_Linear($\underline{x}$)
$\Delta \underline{v} = Y^{-1} \Delta \underline{i}$
foreach node, k

$$dverr_k = |\Delta v_k| - (\kappa_{rel} \cdot v_k + \kappa_{abs})$$

FIGURE 4.5 Algorithm to Calculate DV Errors

4.4.6 Calculation of the Cost Function

Now that we have described all five categories of cost function terms, it is possible to describe the algorithmic flow that is used in their calculation. The terms in $C(\underline{x})$ are calculated from the present problem state in the following order (see Figure 4.6):

1. $devs \leftarrow$ Update__Encapsulated_Devices($\underline{x}$, $\underline{v}$)

   Encapsulated device evaluators are invoked using node voltages and possibly parameters from $\underline{x}$ to determine large-signal device currents and new linearized values for their small-signal models. Again, although $\underline{v}$ can be determined directly from the node voltages in $\underline{x}$, we use distinct notation to make explicit the dependance on the circuit node voltages.

2. $C^{dev} \leftarrow devs$

   The device parameters are used to update any device region-of-operation constraints.

3. $Y \leftarrow$ Stencil_Linearized($devs$), $A \leftarrow$ Stencil_Linearized_Awe($devs$)

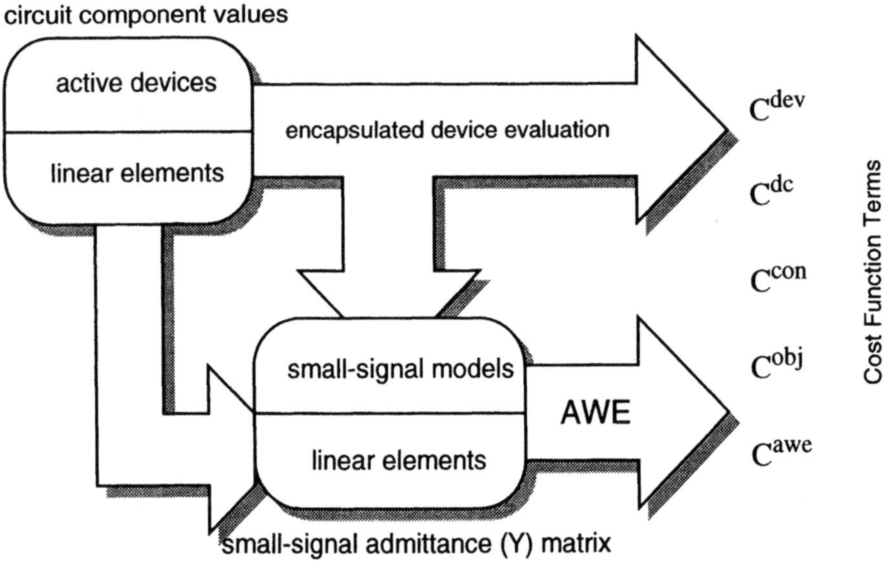

FIGURE 4.6 Evaluation of the Cost Function

The model values are updated in Y for each bias circuit and in A for each AWE circuit.

4. $Y \leftarrow$ Stencil_Linear($\underline{x}$), $A \leftarrow$ Stencil_Linearized_Awe($\underline{x}$)

   The values of any linear elements that depend on $\underline{x}$ are updated in Y for each bias circuit and in A for each AWE circuit.

5. $\underline{H}(s) \leftarrow \mathrm{AWE}(A)$

   AWE is invoked to calculate transfer functions from A. The penalty terms C^{awe} are calculated.

6. $C^{obj} \leftarrow \underline{f}(\underline{x}, devs, \underline{H}(s))$, $C^{con} \leftarrow \underline{g}(\underline{x}, devs, \underline{H}(s))$

   The independent variables, device model parameters, and transfer functions are used to calculate performance characteristics in C^{obj} and C^{con}.

7. $C^{dc} \leftarrow \underline{x}, devs, Y$

   The large-signal error currents from step 4, node voltages from $\underline{x}$, and small-signal matrix Y are used to calculated C^{dc}.

This analysis is extremely fast, allowing a complete re-evaluation for each circuit visited. The cell-level analog circuits we target have less than 100 devices, and produce small-signal models with less than 1000 elements. The bulk of the work is performed by AWE, which reduces such a small-signal model to an accurate low-order transfer function in 10 to 100 ms on a 60 MIP workstation. Previous synthesis strategies were unable to determine both a D.C. operating point and performance characteristics automatically. Our evaluation techniques liberate us from the need to extract from the designer a plethora of performance equations for linear circuit synthesis.

4.4.7 Overview of AWE

Asymptotic Waveform Evaluation (AWE) [Raghavan 93, Pillage 90] is a crucial part of our circuit performance evaluation and thus a crucial part of the calculation of our cost function. In this section, we provide a brief overview of what AWE is and how it works. AWE is itself a subject of much ongoing research, so we present only a somewhat simplified description of the technique and direct the interested reader to [Raghavan 93] to begin more thorough study.

We seek an approximate transfer function, $\hat{H}(s)$, that characterizes the circuit performance of a lumped, linear, time-invariant circuit. We can express $\hat{H}(s)$ as a partial fraction expansion as shown in Equation 4.24.

$$\hat{H}(s) = \sum_{r=1}^{q} \frac{k_r}{s - p_r} \tag{4.24}$$

Here, p_r and k_r are the *r*th *pole* and *residue*, respectively, of the transfer function. Many of the performance metrics of interest to analog circuit designers can be determined from the poles and residues. Taking the sum yields the proper rational form shown in Equation 4.25.

$$\hat{H}(s) = \frac{b_{q-1}s^{q-1} + \ldots + b_1 s + b_0}{a_q s^q + \ldots + a_1 s + a_0} \tag{4.25}$$

Using the form of $\hat{H}(s)$ in Equation 4.25, we can develop the method AWE uses to approximate the actual transfer function, $H(s)$. This method, called Padé approximation, determines values for the coefficients in Equation 4.25. Performing a Maclaurin series expansion of $H(s)$ we can write:

$$H(s) = m_0 + sm_1 + s^2 m_2 + \ldots \tag{4.26}$$

Here, m_i is called the *i*th *moment.*

Setting Equation 4.25 equal to Equation 4.26, cross-multiplying by the denominator, and solving for the denominator coefficients begin the process whereby we can create a set of linear equations of the form:

$$\begin{bmatrix} m_0 & m_1 & \ldots & m_{q-1} \\ m_1 & m_2 & \ldots & m_q \\ \ldots & \ldots & \ldots & \ldots \\ m_q & m_{q+1} & \ldots & m_{2q-1} \end{bmatrix} \begin{bmatrix} a_0 \\ a_1 \\ a_2 \\ a_q \end{bmatrix} = \begin{bmatrix} m_q \\ m_{q+1} \\ \ldots \\ m_{2q} \end{bmatrix} \tag{4.27}$$

We can then solve for the denominator coefficients of Equation 4.25 using standard linear algebra techniques.

Using a root-finder, the poles can be determined from the denominator coefficients. Using these and some additional linear algebra, the residues can then be found.

Approximating $H(s)$ given the circuit moments is only the second half of what AWE must do. First, it must efficiently find the moments given the circuit elements. AWE accomplishes this by iteratively applying a generating matrix. To see this, consider the state equations for a lumped, linear, time-invariant circuit shown in Equation 4.28:

$$\begin{aligned} \dot{x} &= Ax + Bu \\ y &= Cx + Du \end{aligned} \tag{4.28}$$

Here, x is the vector of state variables, u is the input to the circuit, and y is the output from the circuit. Neglecting the direct coupling matrix, D, we can express this as:

$$H(s) = C(sI - A)^{-1}B \tag{4.29}$$

We can then expand $(sI - A)^{-1}$ in a Maclaurin series, as shown in Equation 4.30.

$$\begin{aligned} H(s) &= -C(A^{-1} + sA^{-2} + s^2A^{-3} + \ldots)B \\ H(s) &= -C(A^{-1}B + sA^{-2}B + s^2A^{-3}B + \ldots) \end{aligned} \tag{4.30}$$

The correspondence between Equation 4.30 and the form of $H(s)$ used for the Padé approximation (Equation 4.26) is quite apparent. Thus, all the coefficients needed for the Padé approximation can be calculated by recursive application of the inverted A matrix. In practice, the A matrix used by AWE is formulated in a straightforward fashion from the element values in each of the AWE test jig circuits. An LU factorization can then be performed and the factors used repeatedly.

As previously stated, AWE can be implemented in a very efficient fashion. The compute-intensive tasks in the algorithm are the initial LU factorization of the A matrix, the repeated forward and back substitutions to generate moments, the linear solve to determine the denominator coefficients of $\hat{H}(s)$, the root-finding to determine the poles, and the matrix solve to determine the residues. The size of the matrix used for moment generation is proportional to the size of the circuit being analyzed, while the size of the matrices manipulated during Padé approximation are proportional to the order of the approximation. Using sparse matrix techniques, the algorithmic complexity of the initial LU factorization should be between $O(n^{1.4})$ and $O(n^{1.7})$ [Raghavan 93], depending on the implementation and the circuit. In contrast, the manipulations of the dense matrices used in Padé approximation are $O(n^3)$, where n is the matrix size. For large circuits, the initial sparse LU dominates; however, for smaller circuits the dense matrix operations can dominate for high order approximations. We find that for circuits smaller than about 50 nodes, the dense matrix operations dominate for any approximations of 5th order or higher.

For circuit synthesis, we utilize several important enhancements to the simplified version of the AWE algorithm just described. Two of these are *frequency shifting* and *non-diagonal approximations*. Frequency shifting uses Taylor's series expansions rather than Maclaurin series expansions. Essentially, this means that we approximate around ($s=s_0$), rather than about the origin. This effectively shifts the system poles as shown in Figure 4.7. For stiff systems, where one pole may be near the origin and the next pole far from it, shifting changes the "perspective" of the approximation. The ability to find the second pole in a stiff system is related to the ratio of the distances to the poles from s_0. Thus, as the poles are shifted to the left, as shown in Figure 4.7b, the approximation becomes more numerically stable, and AWE finds the second pole more easily.

The second enhancement important to our work is the ability to find non-diagonal approximations. In our brief discussion of AWE we assumed that the number of poles in the approximation was one more than the number of zeros. However, this assumption is not essential to the derivation and in fact multiple approximations with various numbers of poles and zeros can be determined for the same circuit. Because the Padé approximation has the tendency to occasionally find unstable poles [Chan 91], for example, finding a positive pole in a passive system, we perform multiple approximations and use suitable metrics [Huang 90] to select the best one. Although this increases computation time, AWE is several orders of magnitude faster than conventional techniques, so we can afford it, and this flexibility is essential to effective use of AWE. Typically we explore the available combinations of poles and zeros using one of these methods: a *paradiagonal sequence* or a *horizontal sequence*. A paradiagonal sequence involves approximations with one more pole than zero, and is an efficient method to find accurate, higher order approximations. However, the horizontal

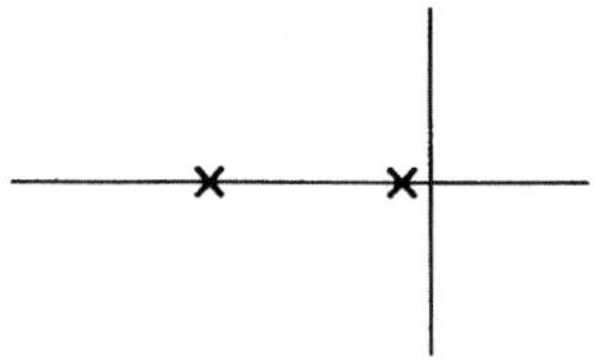

(a) Original Pole Locations

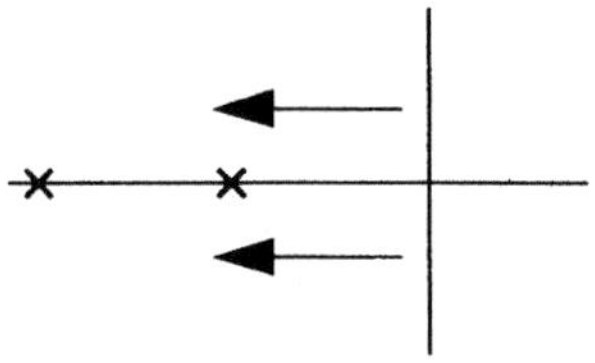

(b) Shifted Pole Locations

FIGURE 4.7 Frequency Shifting in AWE

sequence, which fixes the number of poles and varies the number of zeros, is guaranteed to converge to the system's true poles. In practice, we typically select the paradiagonal sequence because of its efficiency.

4.4.8 Computing Performance Metrics with AWE

Once an AWE approximation for a circuit ($\hat{H}(s)$) has been calculated, it is used to calculate performance metrics of interest to the user. Performance objective and constraint functions specified by the user can make use of predefined functions that manipulate AWE transfer functions and return performance information such as gain and bandwidth. The complete list of predefined functions that use AWE is given in Table 4.1 .

Most of these are algorithmically straightforward. However, a few algorithmic details are worth discussing. First, the AWE approximations are performed on demand and cached for each circuit. Thus, a predefined function will request an AWE approximation and if the appropriate approximation is not available it is performed. Since all the approximations for a given circuit and shift frequency can be performed from a single LU factorization for that circuit, the cache provides substantial CPU savings.

Another algorithmic point of interest is that several approximations with different shift frequencies may be required for a single predefined function evaluation. For example, the unity_gain_freq function is optimized for a transfer function with a low frequency dominant pole, such as that of most operational amplifiers. To determine this pole, an unshifted approximation is performed. (Because shifted and unshifted approximations require slightly different Y matrices, two separate matrices are kept. This approach optimizes execution time over memory.) The D.C. gain and dominant pole are extracted from the unshifted approximation. A dominant pole approximation predicts that the unity gain frequency is simply the product of the gain and pole. This initial guess is used as a shift frequency to generate a new approximation that is more accurate near the unity gain frequency. The unity gain frequency from this shifted approximation is then calculated. If the new unity gain frequency is significantly different from the old unity gain frequency, another shifted approximation is performed. The new shift frequency is computed as the geometric mean of the old shift frequency and the new unity gain frequency. This cycle of calculating the unity gain frequency and updating the shift frequency is repeated until the unity gain frequency converges. We have found this technique to converge quickly in practice.

In this section we described in detail the five components of the cost function and how they are calculated. The remaining component that must be described to com-

TABLE 4.1 Predefined Functions that use AWE

Function	Other Arguments	Description
unity_gain_freq		The unity gain frequency; the frequency at which $\lvert\hat{H}(s)\rvert = 1$.
phase_margin		The phase difference between π rad and the phase at the unity gain frequency.
phase	frequency	$\angle(\hat{H}(\text{frequency}))$
gain	frequency	$\lvert\hat{H}(\text{frequency})\rvert$
dc_gain		$\lvert\hat{H}(0)\rvert$
min_gain	f_{min}, f_{max}	$\min_{f_{min} \le f \le f_{max}} \lvert H(f)\rvert$
max_gain	f_{min}, f_{max}	$\max_{f_{min} \le f \le f_{max}} \lvert H(f)\rvert$
settle_time	vin, e_{abs}, e_{rel}, *guess*	Given that $\hat{H}(s)$ is excited by a step input of height vin, compute the time taken to settle within e_{abs} absolute error or e_{rel} relative error from the final answer. The *guess* is used to select an appropriate shift frequency.
max_pole_real	f_{guess}	Return the real part of the pole with the largest magnitude. f_{guess} is used to select an appropriate shift frequency.
max_pole_imag	f_{guess}	Return the imaginary part of the pole with the largest magnitude. f_{guess} is used to select an appropriate shift frequency.
max_pole_residue	f_{guess}	Return the residue associated with the pole with the largest magnitude. f_{guess} is used to select an appropriate shift frequency.

plete the annealing formulation is the set of annealing control mechanisms, such as the cooling schedule.

4.5 Annealing Control Mechanisms

The fourth and final aspect of our annealing design is the set of control mechanisms used to make the optimization as efficient and robust as possible. In this section we present the chief components of our control mechanism: the cooling schedule and the weight control mechanism.

4.5.1 Cooling Schedule

As discussed in Section 4.1, the cooling schedule defines the initial starting temperature and the functions *done_at_temperature, update_temp* and *frozen*. For temperature control—which defines *done_at_temperature* and *update_temp*—we have implemented the general-purpose annealing schedule of Lam [Lam 88, Lam 88a, Lam 86] as modified by [Swartz 90]. Our freezing criteria, which determines when the annealing is complete has been developed specifically for our analog synthesis application.

The modified Lam schedule

The general cooling schedule of Lam, implements temperature control in *update_temp* using a fairly complex strategy, presented in [Lam 88]. Initially we implemented a different general-purpose cooling schedule [Huang 86], but switched to the Lam schedule to reduce run-time. Comparisons between the Lam schedule and other schedules, including the one in [Huang 86], show that the Lam schedule performs better for most applications. More importantly, the Lam schedule allows greater control over the total number of moves the annealer will perform. This greater control is the foundation upon which our dynamic weight mechanism is built (see Section 4.5.2).

The derivation of the Lam schedule, also referred to as the "efficient λ-schedule", is quite complex. Its premise is that annealing is most successful when the system being annealed is kept close to thermal equilibrium as the temperature is lowered. Because this would require too much computation time to be practical, useful annealing implementations must compromise between the quality of the final solution and the computation time. Lam defines a control parameter, λ, to quantify this trade-off. The goal of the efficient λ-schedule is to keep the problem in *quasi-equilibrium*, where λ controls how well this approximates real equilibrium. To do this, the temperature is updated using a formula that incorporates λ, the sampled mean and standard deviation of the cost function over a sliding window of recent moves, and the accept ratio (the ratio of accepted to generated moves). In theory, the temperature should be updated after each move, eliminating the need for *done_at_temperature* (see Figure 4.1), but because of the computational complexity of the update formula, practical implementations usually update the temperature every 5 or 10 moves.

Results for the Lam schedule are quite impressive. When compared with other general-purpose annealing schedules such as [Huang 86] and even hand-crafted schedules like the one in [Sechen 86], it often provides speed-ups of 50% while actually improving the quality of the final answers [Lam 88a]. Simulated annealing with the Lam schedule even compares favorably with heuristic combinatorial optimization methods tuned to specific problems like partitioning and the travelling salesman problem [Lam 88].

Unfortunately, the original Lam schedule requires the use of a move range control scheme that exponentially decreases the allowable range to control the accept ratio. As discussed in Section 4.3 this simple move control scheme is unacceptable for the large number of possible moves for circuit synthesis problems of reasonable size. Because of this incompatibility with our move control scheme and also because of the complexity of the Lam schedule, we implement a modification of it due to Swartz [Swartz 90]. The use of the modified Lam schedule allowed us to use our existing move control mechanism with only minor changes, detailed at the end of this section. The modified Lam schedule is based on empirical observation of a large number of annealing runs using the full Lam schedule. If the accept rate is plotted with respect to number of moves, these runs have an almost identical pattern, shown in Figure 4.8. The accept rate starts at about 100%, decreases exponentially until it stabilizes (about 15% of the way through the run) at about 44%, and remains there until about 65% of the way through the run. It then continues its exponential decline until the end of the annealing process. Based on this observation, Swartz duplicated the effect of the Lam schedule by using a simple feedback loop to control the temperature and *force* the accept ratio to follow the curve in Figure 4.8. In comparing the modified schedule to the original schedule, Swartz reported almost no difference in the quality of the final answers [Swartz 90]. One additional benefit of the modified Lam schedule is that, in contrast to the original schedule, the total number of moves in the annealing run can be specified directly by the user. This also implies that the annealer always knows how far along in the annealing process it is. This ability is extremely important to the dynamic weight control mechanism described in Section 4.5.2.

Until now we have neglected an important aspect of both cooling schedules: the interaction between the move-set and the cooling schedule. Even in the modified Lam schedule, the range limits for $\Delta \underline{x}$ are controlled as part of the cooling schedule itself. The ranges decrease exponentially as a function of the number of moves generated—not as a function of temperature. Since the temperature increases or decreases as necessary to force the accept rate to follow the profile of Figure 4.8, it is the shrinking range limits that eventually force the problem to freeze. As the allowable moves become smaller, they are more readily accepted and the accept ratio has a tendency to rise. To compensate for the rising accept rate, the temperature will fall, and eventually the problem will freeze. In our annealing formulation, the move-set is separately controlled, as described in Section 4.3. Because our method increases a move class's selection probability when it has a large quality factor, it tends to favor large moves that change the cost function substantially. Thus, neither the temperature nor the move selection mechanism push the problem towards a frozen state. To fix this problem, we employ a hybrid move control scheme that adds exponential decay to the automatic move selection mechanism of Section 4.3. After a certain percentage of the total moves, ρ, have been performed, we modify the move classes of Section 4.3, exponentially decreasing the size of *each* move range, *r*. For example, suppose we

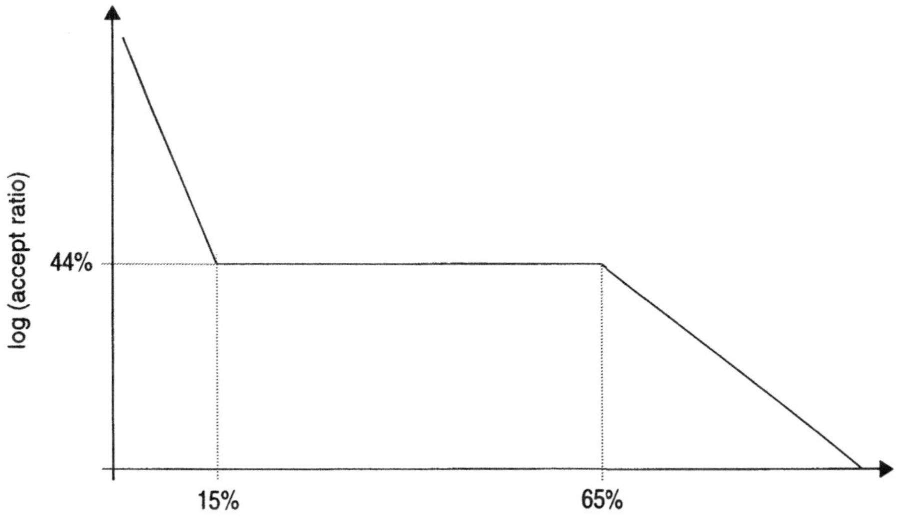

FIGURE 4.8 Accept Rate Target for the Modified Lam Schedule

plan to try 100,000 moves and set ρ to 40%. For the first 40,000 moves the ranges of the move classes are unchanged, and move class i perturbs x_i with $[r^{min}, r^{max}]$. However, after 40,000 moves, the range $[r^{min}, r^{max}]$ decreases exponentially. The rate of decrease varies from move class to move class, and is set such that for a given variable all move ranges end up as small as the smallest move range at the beginning of the anneal. This exponential reduction has the effect of more quickly forcing the problem state to freeze. The percentage of moves we must perform before beginning this shrinking process ρ must not be too large or the process will freeze too late, but shrinking the moves too early will lead to poorer solutions on average. The selection of the best value for this parameter is discussed in detail in Section 7.1.

Estimating the Number of Moves to Attempt

A final aspect of the modified Lam schedule is determining the number of moves to make for a given circuit problem. As a simple default, we use a formula based on work in [Sechen 87], in which Sechen argues that the number of moves required for effective annealing increases as:

$$K \times N^{\frac{4}{3}} \tag{4.31}$$

where N is a measure of the problem size and K is a constant. As we shall see in Section 6.5.1, for our synthesis problems, an accurate measure of problem size is difficult to derive. As an initial estimate, our annealer uses the number of independent variables, although this can easily be overridden. The value of K was determined experimentally as discussed in Section 7.1. Although simple, this mechanism has proven quite robust.

The Freezing Criteria

Our freezing criteria is somewhat unusual because our problem state combines both discrete and continuous variables. Conceptually, the task is simple. After the minimum number of moves required for the Lam schedule has been performed, we consider the problem state, $\underline{x}$, to be frozen if it has stopped changing, or nearly stopped changing. To quantify this concept, we sample the problem state every G generated moves, comparing the present state $\underline{x}_{G_0}$ with N states examined G moves previously: $\underline{x}_{G_{-1}}, \ldots, \underline{x}_{G_{-N}}$. Then, for the ith term in $\underline{x}$ we define:

$$min_{x_i} = \underset{k \in [-N, 0]}{\text{minimum}} \; x_{G_k i} \tag{4.32}$$

and

$$max_{x_i} = \underset{k \in [-N, 0]}{\text{maximum}} \; x_{G_k i} \quad \textbf{(4.33)}$$

Thus, min_{x_i} is an estimate of the smallest value for x_i over the previous $N \times G$ moves, and max_{x_i} is an estimate of the largest value for x_i over the same interval. We then consider the problem frozen if the estimated change in the variable over the sampling interval is within some relative tolerance, ε. This is formalized in Equation 4.34.

$$\frac{max_{x_i} - min_{x_i}}{max_{x_i}} < \varepsilon_i, \forall (x_i \in \underline{x}) \quad \textbf{(4.34)}$$

For discrete variables, we use integer indices into the grid of allowable values for this calculation and ε_i is always 1. This is equivalent to checking that the value has not changed over the previous N samples. For real variables, the relative tolerance, ε_i is determined from KCL constants and other constraints in the problem description. This freezing criteria, coupled with the modified Lam cooling schedule, completely define the *done_at_temperature, update_temp* and *frozen* functions of Figure 4.1 on page 82. What remains to complete the annealing formulation is to specify how the weights in the cost function are determined.

4.5.2 Dynamic Weights

One of the most serious drawbacks of using simulated annealing is the difficulty of balancing the terms in the cost function. This is exacerbated by the mapping of a constrained optimization to an unconstrained one, but is also a result of the large number of interacting objectives in the analog synthesis problem. This balancing process is usually accomplished by introducing weights, w_i, into each of the terms of the cost function, as we have seen in Section 4.4. The problem then becomes one of selecting the appropriate weights to achieve the desired answer. We combat this problem in two ways: by providing the term-wise normalization functions of Section 4.4 and by using algorithmic mechanisms to dynamically set the weights.

In early tests of our annealing formulation, the need for more sophisticated control of the weights became evident. For example, if the weights used in the C^{dc} terms were too small, the final solution was not D.C. correct, so its performance prediction did not match simulation results. If the weights in the C^{dc} term were too large, the annealer did not explore the design space sufficiently, resulting in poor quality answers. Similar behavior was observed with the weights of the other penalty terms. A reasonable explanation for this is that the solution for a constrained optimization problem typically lies on the edge of the feasible region. When the weight on a penalty term is too large, it forces the search too far into the feasible region and the best solution cannot be found. If the weight is too small, the search ends up too far from the feasible region. The ideal weight should control the contribution of each of the penalty terms so that the bulk of the search is conducted in a region very near the edge of the feasible region, forcing them to be feasible just before the final solution was found. Unfortunately, it was found in early experiments that to increase the probability of converging to a feasible solution required a fixed weight larger than the weight values that produced the best objective values (but poor convergence). To combat this problem, we designed a dynamic weight system that adjusts the weights algorithmically. Although we use a unique formulation, the idea of non-fixed weights is not new to simulated annealing, *e.g.*, a similar formulation is in [Sechen 87]. Conceptually, we would like to keep the search near the feasible region during the bulk of the optimization process and force convergence before the final freezing of the problem state. This algorithmic weight scheme has the added advantage that it is more robust in the face of a wide variety of circuits, independent variables, and performance metrics than fixed weight schemes, which must be adjusted by a human to compensate for changing problem specifications. Our dynamic weight scheme is based on the following observations:

1. The range and tolerance of the penalty terms are relatively easy to determine. For example, for the C^{dc} terms the tolerances τ_{rel} and τ_{abs} define how small the Kirchhoff's law errors must be for the circuit to be considered D.C. correct. We can use this to estimate how close to feasible we must be to have a high probability of final convergence.
2. The relative contribution of a term to the overall cost function determines the amount of effort the annealer expends to reduce that term. If a term's contribution is negligible with respect to the overall cost, it will be "in the noise" and fluctuate almost randomly. The weights can be used to increase the relative contribution of a particular term to the overall cost, thereby forcing the annealer to "work harder" to reduce the error introduced by that term.
3. Because we are using the modified Lam cooling schedule, we know approximately *where* we are in the overall optimization process. Thus we can predict *when* we should change our focus from remaining close to the feasible region to forcing convergence.

Using these three observations, we implement a dynamic weight system with two components: a *trajectory* that the error associated with each penalty term should follow, and a *feedback control system* that balances the relative demands of all the penalty terms with the objective terms.

There are two broad classes of error trajectories: one for D.C. error terms and one for the other kinds of terms. The general shape of both trajectories is shown in Figure 4.9. Early in the annealing process, the D.C. term errors are forced to be as small as possible. This will result in large weights on those terms to force the annealer into a region of the cost function where the C^{DC} terms are small. During this time, the model and performance constraints are relaxed; however, once C^{DC} is small, the allowable errors on the performance constraints are tightened. This sequential introduction of constraints is similar to the method of introducing constraints used in [Maulik 92a]. For the majority of the optimization, the error requirements are kept stable, forcing the explored states to be near enough to being met that the objective terms can be minimized. Finally, during the final stages of cooling, the error requirements are tightened, forcing all the penalty terms to be met absolutely in the final solution. Note that the exact shape of these four-segment piecewise linear trajectories is completely specified by the four points p0, p1, p2, and p3. The exact locations of these points and how they were set for the various types of penalty terms is discussed in Section 7.1.

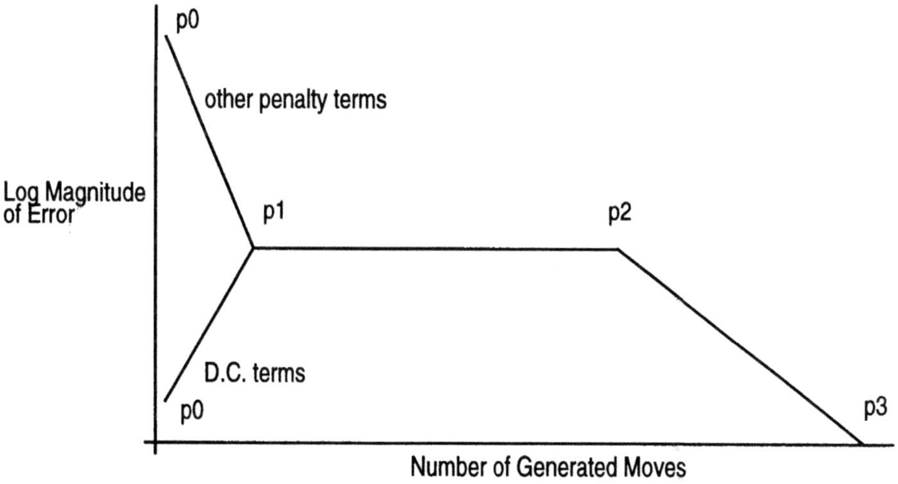

FIGURE 4.9 General Trajectory That the Error for a Penalty Term Should Follow

The second component of the dynamic weight system is the feedback control mechanism, which is itself implemented in two phases: a per-term phase, shown in Figure 4.10, and an overall phase, shown in Figure 4.11. The dynamic weight mechanism updates the weights for each term a fixed number of times over the anneal. The computations are based on average costs and errors over a sliding window that typically covers the last few weight updates. This window is called a *statistical interval.* Note that since there are no weights on C^{OBJ}, the other terms use this term as a reference value. As shown in Figure 4.10, the per-term phase increases the weight when the error over the last statistical interval is much larger than the error trajectory value for the present phase of the anneal. This will increase the contribution of this term to the overall cost function and the annealing action will then reduce the error faster.

if (error is greater than trajectory)
    if (error is decreasing)
        no change
    else (error is increasing)

$$w = w\left(1 + K\frac{\text{error} - \text{trajectory}}{\max(\text{error}, \text{trajectory})}\right)$$

else (error is less than target)
    if (cost is less than target term)
        no change
    else (cost is greater than target term)

$$w = w\left(1 - K\frac{\text{cost} - C^{OBJ}}{\max(\text{cost}, C^{OBJ})}\right)$$

FIGURE 4.10 Algorithm for Per-Term Dynamic Weight Adjustment

old_cost = $C(\underline{x}) - C^{OBJ}$, with old weights
new_cost = $C(\underline{x}) - C^{OBJ}$, with new weights.
if (new_cost > old_cost)
    for each weight, w_i

$$w_i = w_i\left(1 - K\frac{\text{new\_cost} - \text{old\_cost}}{\text{new\_cost}}\right)$$

FIGURE 4.11 Algorithm for Overall Dynamic Weight Adjustment

Conversely, when the error is much smaller than the required value, the weight can be decreased, allowing the annealer to focus on other terms in the cost function. As shown in Figure 4.11, the over-all phase compares the value of the overall cost function before and after the per-term adjustments, re-normalizing all the weights to ensure that there is no net change in the cost function value.

The dynamic weight mechanism is the last aspect of the control mechanisms component of our annealing formulation. In the previous four sections of this chapter, we have described completely our annealing formulation via its four components: the problem representation, the annealing move-set, the cost function, and the annealing control mechanisms. In the next section we provide a brief overview of the implementation of our annealing formulation in the tool OBLX.

4.6 Implementation Details

The annealing formulation for analog synthesis discussed in this chapter has been implemented as a library of code called OBLX. This library is linked with the circuit specific components of the cost function which are automatically-generated in code

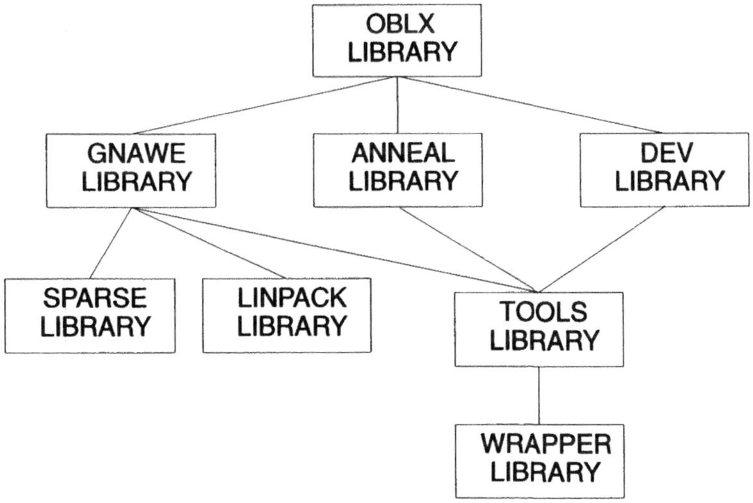

FIGURE 4.12 Software structure of the OBLX library

form by the companion tool ASTRX, discussed in the next chapter. The result is an executable, automatically generated circuit synthesis program specifically tailored to design the circuit specified in the problem description. Running this program will yield a single design of the circuit to match the given specifications.

The OBLX library itself is built on several layers of software as shown in Figure 4.12. The description of these various component libraries is contained in Table 4.2 . Com-

TABLE 4.2 Component Libraries of OBLX

Name	Description	Origin	Source Lines
OBLX	General annealing structure to link to circuit specific routines generated by ASTRX. Includes initialization, move generation, Newton-Raphson, and generalizable terms in the cost function.	for OBLX	8000
GNAWE	A re-implementation of AWE optimized for repetitive use.	for OBLX, & RAIL [Stanisic 96]	12700
Anneal	A general purpose annealing tool-kit that provides a convenient way to compare several cooling strategies, dynamic weight schemes and other aspects of simulated annealing.	for OBLX	12000
DEV	Non-linear circuit device models adapted from SPICE. Presently includes BSIM and MOS3 models for MOSFETs and Gummel-Poon model for BJTs.	SPICE3/ for OBLX	26200/ 3000
Sparse	A sparse matrix LU factorization library.	Berkeley	13000
Linpack	Linear Algebra library		6500
Tools	Programmer's Tool-kit. Portions used include: dynamic arrays, hash tables, lexical analyzer, execution manager, complex number routines, error manager, polynomial routines, random number generator, initialization file manager, resource manager, and statistical tool-kit.	for ASTRX/ OBLX	13300
Wrapper	A portability layer built on the UNIX library.	local	3900
	Total / Total local		98600/ 52900

plete discussion of tests performed and implementation results is contained in Chapter 6 and Chapter 7.

4.7 Summary

In this chapter we presented the details of the optimization formulation we use to solve the nominal analog circuit synthesis problem. We expended substantial effort designing the components of the simulated annealing algorithm, both in adapting it for our purposes and in making it as efficient and robust as possible. Because of the mix of voltage variables and discrete device sizes, we employ a mix of continuous and discrete variables within the annealer. The annealer can select among a palette of several move sizes for each variable, as well as moves that use the Newton-Raphson algorithm to move towards D.C. correctness. A dynamic move class control mechanism uses a move's recent performance to bias the probability that a given class of moves will be selected. The cost function used by the annealer combines user supplied performance metrics, device region constraints, and relaxed D.C. terms. The annealer also uses sophisticated control mechanisms such as the Lam cooling schedule and a dynamic weight adjustment mechanism to ensure efficiency and robustness. Finally, the annealer is implemented as a suite of libraries that build upon each other to provide maximum code flexibility, reusability and maintainability.

CHAPTER 5 A Circuit Compiler

In the previous chapter, we described the algorithms used to minimize a cost function, $C(\underline{x})$, carefully constructed so that its minimum value occurs when $\underline{x}$, the component and voltage values in the circuit, specifies an optimal design of the circuit to synthesize. In this chapter, we describe the algorithms used to compile the circuit description given by the user into code that evaluates $C(\underline{x})$. We first discuss the problem description generally, and the information the user must provide. We then discuss the algorithms used to analyze this description. Finally, we discuss the code generation process itself.

5.1 Required Input

We refer to the input provided by the user to the circuit synthesis problem as a single monolithic object called the *problem description*, which contains all the information needed to complete the synthesis task. In Section 3.4 we presented a simple example of a synthesis task and discussed informally the contents of the problem description. In this section, we present a more formal description of those contents. To make our discussion concrete, we again use the simple synthesis problem discussed in Section 3.4. For convenience, we summarize this example in Figure 5.1.

Conceptually, the typical contents of the problem description are shown in Figure 5.2 and consist of the following eight components:

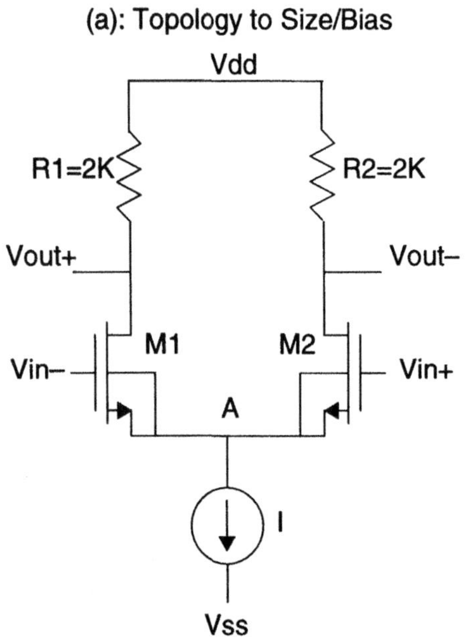

(b): Performance Specifications

Attribute	Specification
A_{dm}	maximize
UGF	≥ 1MHz
A_{cm}	≤ 0dB
SR	≥ 1V/μs

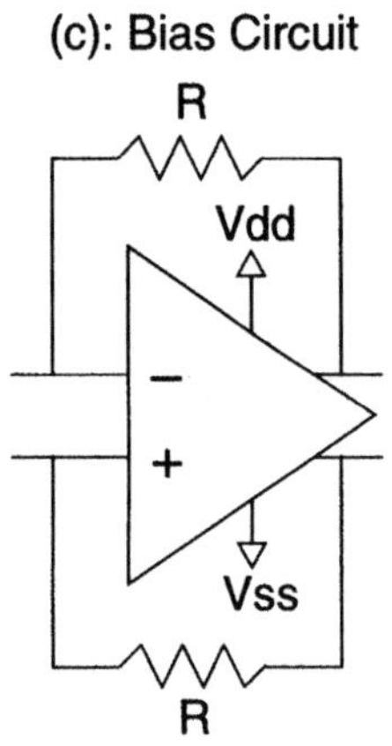

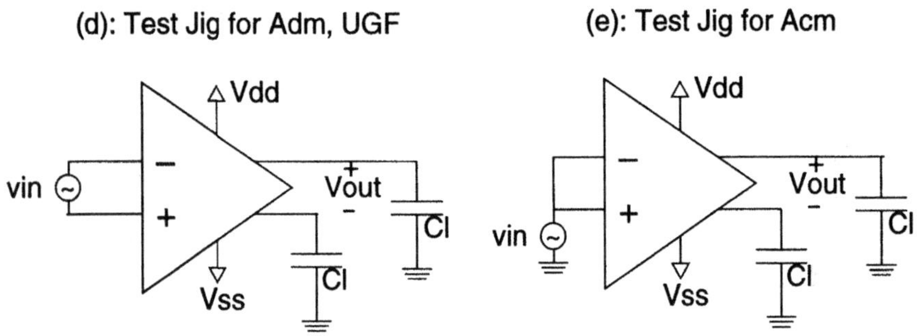

FIGURE 5.1 A Design Example: Input Required to Complete a Circuit Synthesis Problem Description.

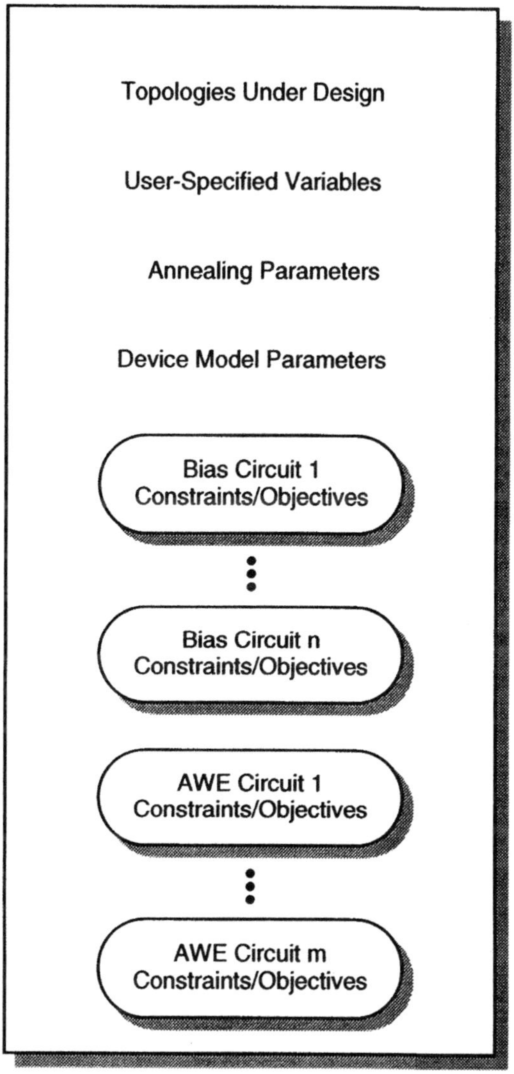

FIGURE 5.2 Eight Components of the Input Description

1. **Topologies under design**. The first part of the description is/are the topology/topologies of the circuit/circuits to be sized and biased. In our example, we are designing a single circuit; however, in a more complex application we may wish to design *simultaneously* a set of related circuits or circuit components. The topologies are described as independent sub-circuits that can then be included in test jigs later in the description.
2. **User-specified variables**. These form part of the set of independent variables, $\underline{x}$, perturbed by the annealer during the solution process. Recall that because of the relaxed D.C. formulation, the annealer also manipulates node voltages as independent variables in order to satisfy Kirchhoff's laws.

   Analog designers know that simple relationships between device values must be maintained for certain circuits. For example, in a symmetric circuit the component values in one half must match the component values in the other. Thus, it is not enough to provide a distinct variable for every component whose value must be determined by the synthesis tool. Instead, as introduced in Section 4.2, we choose to allow the designer to specify a set of *independent* variables and then allow the values of components to be simple expressions over these variables. Circuit element values or device parameters in the topologies that are unknown are specified as expressions that contain one or more of the user-specified variables. If the value is fixed, the expression is a simple constant. This simple expression mechanism gives the designer substantial flexibility in defining relationships between component values. For example, it is simple to specify that the length of one bias transistor must be twice the length of another. If the expression for one transistor's length is 'L', then the second transistor's is simply '$2 \times L$'.
3. **Annealing parameters**. The user can override any of the constants used during the annealing process. This includes all the controls over the cooling schedule, weights, and tolerances such as τ_{abs}, the absolute tolerance on D.C. convergence. Typically, overriding the defaults will not be necessary. For our example, and all the circuit results in this book, we do not change the defaults.
4. **Model parameters for active devices**. This information parameterizes the appropriate encapsulated device evaluators, allowing them to determine the behavior of the devices in the circuit. The parameters are derived based on an analysis of the IC fabrication process that will be used to create the device. Different evaluators, such as MOS3 and BSIM have their own set of required parameters. Typically, the parameters are just a rather large set of numbers.
5. **Set of bias circuits**. Since AWE works on linear circuits only, non-linear circuits are linearized automatically to equivalent small-signal (*i.e.*, linear) models. This linearization is the same approach used in simulators like SPICE [Nagel 75]. Recall that the small-signal models are generated by the encapsulated devices, but

require operating point information (a D.C. bias point) to generate the appropriate model. To generate this operating point information, the user specifies one or more bias circuits. Typically, these circuits contain instances of the circuit topologies under design, power supplies, and other bias circuitry. For the example in Figure 5.1, the bias circuit is shown in Figure 5.1(c) and includes an instance of the amplifier, power supplies, and a feedback path from output to input.

6. **Set of AWE circuits**. These are the test jigs, which provide circuit conditions under which to test the performance of the circuit to synthesize. They typically include instances of the circuit to synthesize and any additional components for loading, *etc.* For our example, to test the open-loop characteristics of our amplifier, A_{dm} and UGF, we use the circuit on the lower left in Figure 5.1. This AWE circuit includes an instance of the *amp*, power supplies, the input signal, and load capacitances. A second AWE circuit (on the lower right of Figure 5.1) is necessary to test for the common mode gain of the circuit.

   Recall that AWE computes transfer functions from an input source to an output node. The performance of interest is then derived from the transfer function. Thus, AWE circuits will include one or more transfer function definitions, each with a name, source and output node. Because transfer functions are often processed further via predefined functions, each must be assigned a unique name.

   Each AWE circuit must also be associated with a bias circuit. Recall that non-linear devices are allowed in AWE circuits, but as with any simulator, inside the annealer, before being passed to AWE each device is replaced with a linearized small-signal model. To generate the small-signal model, an encapsulated device evaluator requires not only the device's physical dimensions, but its port voltages as well. In Section 3.4 we discussed our decision to use a separate circuit with which to determine these bias voltages. Thus, these port voltages are actually taken from the bias circuit associated with the AWE circuit. Each non-linear device in an AWE circuit must be *linked* to a non-linear device in a bias circuit. A small-signal model is derived from the linked device (in the bias circuit) and replaces the device in the AWE circuit. To simplify this mechanism we design the compiler such that these bias/AWE device links are made automatically by matching device names. However, to provide complete flexibility, we have also included a mechanism whereby the user can override this by-name matching and explicitly specify any bias/AWE device link.

7. **Objective/Constraint functions**. These correspond to C^{obj} and C^{con} in the cost function. Typical objective functions are performance characteristics such as gain, active device area, or static power consumption. The function specifications may use built-in functions that derive performance characteristics such as gain from the AWE transfer functions specified in 6 (above). For our example, this mechanism

makes it simple to specify A_{dm}, A_{cm}, and UGF. Function specifications may also use circuit values such as node voltages and currents. This capability is necessary to write simple performance equations, such as the slew rate equation for our example. We discuss the full expression capabilities of the compiler in Section 5.1.1.

8. **Constants, functions, and expressions.** To further increase the ease of describing a circuit synthesis problem, we design the input description such that the user may easily define constants and functions. These user-defined functions provide a simple but powerful method for defining more complex specifications. A function or constant may be used anywhere a value is expected. Where possible, the compiler interprets functions and resolves them to constants. If the function involves user-specified independent variables, the compiler generates code so that the function can be evaluated during the annealing process.

In the remainder of this section, we first discuss in more detail the design and use of the function and expression capability built into the compiler. We then present some details of the actual look and format used for the input description and present an example.

5.1.1 Constants, Functions and Expressions

One of the most important aspects of the input description design is the capability to handle constants, functions, and expressions. As with a programming language, these provide the ability to abstract repeated sections of the input, replacing many instances with a single generic pattern. However, for our synthesis problem, expressions also provide the essential ability to write equations that are compiled and included in the cost function used during annealing. We find that constants are often convenient in a problem description to specify parameters that may change. For example, replacing the values of the power rails with the symbolic constants vddval and vssval provides additional documentation and makes the values convenient to change. This same characteristic holds true of functions, and, for our rudimentary functional language, the definition of a function is given by writing an expression. However, expressions are also the mechanism used to describe performance objectives, performance constraints, and device region constraints that will be compiled into the annealer's cost function. In our synthesis example, our performance objective is to maximize A_{dm}, which we specify as the objective function $f_{Adm}(\underline{x})$. In the input description, we can write the expression $f_{Adm}(\underline{x})$ as:

$$\text{dc\_gain(tf)} \quad (5.1)$$

where tf refers to an AWE transfer function defined as Vout+/Vin for the test jig on the lower left of Figure 5.1. Using the notation from Section 4.4.1, $f_{Adm}(\underline{x})$ is the only member of the set of objective functions, Ω^{OBJ}, and we need to map this to the objective term in the cost function, $C^{obj}(\underline{x})$. Recall that,

$$C^{obj}(\underline{x}) = \frac{1}{\|\Omega^{obj}\|} \cdot \sum_{i \in \Omega^{obj}} \hat{f}_i(\underline{x}) \tag{5.2}$$

where,

$$\hat{f}_i(\underline{x}) = \frac{f_i(\underline{x}) - good_i}{bad_i - good_i} \tag{5.3}$$

Since $C^{obj}(\underline{x})$ is a function of $\underline{x}$, its value cannot be determined by the compiler. Instead, the compiler analyzes the expression for $f_{Adm}(\underline{x})$ provided by the user and writes code to evaluate Equation 5.2 and Equation 5.3. This code is used by the annealer to re-evaluate $C^{obj}(\underline{x})$ for each new circuit configuration visited by the annealer.

The syntax of these expressions is designed to be simple yet powerful. The basic syntax is shown in Figure 5.3. This description formalizes an expression syntax that is similar to that in the C programming language. The expressions also rely a great deal on predefined functions and variables. Since we are trying to automate as much as possible the circuit synthesis process, these predefined functions and variables are intended to make all the available information about the circuit under design readily accessible to the designer writing the input description. The predefined functions can be grouped into three categories: math functions, AWE functions, and circuit functions. For completeness, the math functions supported are listed in Table 5.1 . The predefined functions that manipulate AWE transfer functions were discussed in Section 4.4.8 on page 105. And the other circuit functions are listed in Table 5.2 . The predefined variables are used to access circuit information such as the values of small-signal elements in a non-linear device model or the voltage at a particular node. Because the annealer fills in the actual value of the variable for each new circuit configuration, the analog designer can use these variables to write very accurate performance equations. The actual names of available predefined variables vary based on the circuit and encapsulated devices in use. For each active device, all the parameters returned by the encapsulated model can be used in expressions. For example, the variable `xamp.M1.vdsat` refers to the saturation voltage for a transistor M1 in a circuit named xamp. When a variable of this form is encountered, the compiler interrogates the encapsulated evaluator to see if the variable name is valid. If so, the variable is replaced with code that links it to the appropriate parameter. During annealing, these links are followed so that the value is always up-to-date. The large number of predefined variables and functions supported by the compiler make writing small performance equations straightforward. More complex performance metrics can be built from these writing other functions. Although this expression capability is not as general as a full programming language or a hardware description language, it is powerful enough to describe the synthesis task in a fashion that is easy to use and familiar to most analog designers.

5.1.2 Look and Format of the Input Description

Thus far, we have described conceptually the components in the input description; however, there is a fair amount of flexibility in the format used to actually write this information in a file. Here, our chief goal in designing the actual look and format of

```
expr        ←   relation
            |   relation '?' expr ':' expr

relation    ←   relation rel_op relation
            |   math

math        ←   math op math
            |   simple

simple      ←   NAME '(' arguments ')' function call
            |   NAME                   variable reference
            |   NUMBER
            |   '(' expr ')'
            |   '-' simple

arguments   ←   expr ',' arguments
            |   expr

rel_op      ←   '&&' | '||' | '==' | '!='
            |   '<' | '<=' | '>' | '>='

op          ←   '+' | '-' | '*' | '/'
```

FIGURE 5.3 Expression Syntax Supported by the Compiler

the input description *file* was to abide by the principal of "least amazement" [Wirth 74], which states that things that are not new should not appear to be new. Since analog designers already work with computer representations of circuit topologies and many of the other parts of the input description, we choose to reuse, as much as possible, existing formats that are familiar to analog designers. We could have based our input description on one of the evolving hardware description languages for analog circuits [Cohen 88, Intermetrics 88, Saleh 94]. Although these new languages are powerful enough to describe our synthesis problem, because of their complexity learning them would be a substantial barrier to learning ASTRX/OBLX. Instead,

TABLE 5.1 Predefined Mathematical Functions

Name	Description	Name	Description
sin	Natural Sine	sinh	Hyperbolic Sine
cos	Natural Cosine	cosh	Hyperbolic Cosine
tan	Natural Tangent	tanh	Hyperbolic Tangent
atan	Arc-Tangent	sqrt	Square root
exp	Natural Exponent	db	20 * log10
log	Natural Logarithm	log10	Logarithm Base 10
abs	Absolute Value	fabs	Absolute Value
min	Minimum (2 arguments)	max	Maximum (2 arguments)

TABLE 5.2 Circuit Related Predefined Functions

Name	Description
dev_area()	The area of the active devices. This is calculated by the encapsulated device evaluators. However, this is still an estimate and actual layout dimensions may vary.
cap_area()	The area used by capacitors. This is an estimate calculated from the capacitance per unit area for the process.
dev_power()	Power used by the active devices. Calculated by encapsulated device evaluators.
res_power()	Power dissipated by resistances and conductances in the circuit.
v(node)	The voltage at the given node.

because analog designers are already familiar with detailed circuit simulators, the input format is designed to resemble that of SPICE [Nagel 75]—one of the most commonplace circuit simulators—as closely as possible. Actually, the syntax was patterned after a commercial implementation of SPICE, called HSPICE [Meta 87]. HSPICE was selected for a number of reasons. First, in most respects it is the same as SPICE and thus widely known. Second, it is available locally and documentation of the input format is readily available and more complete than that of SPICE. Third, many of the enhancements needed for synthesis have already been included in the HSPICE language. Thus, the format for the constructs needed for the input description file was based as much as possible on that of HSPICE, to make the format look as familiar as possible to designers. This lowers the learning barrier for designers new to the tool.

5.1.3 Input Description for the Design Example

The input description file for the example in Figure 5.1 is shown in Figure 5.4. The labels on the left side of the figure indicate the logical components of the file. Note that the AWE transfer functions are lines that begin with '.pz'. Each transfer function is a (node, source) pair. For example, in line 39 of Figure 5.4, 'tf' is defined as the transfer function between the input voltage source and the output node.

5.2 Analysis

After validation of the input, the problem description is analyzed. All the problem-specific information needed for annealing is derived during this analysis phase. We partition the compilation problem so that code generation merely prints out information already determined by analysis. We do this because much of the code generated by the compiler relies on the same pool of derived information. Further, the generated code must be completely consistent. By deriving everything we need in advance, we avoid repeated analyses during code generation and inconsistencies in the generated code. In fact, most of the code to be generated is produced and stored as strings during analysis. A high-level pseudo-code description of the analysis consists of the steps shown in Figure 5.5.

The majority of these steps require further explanation. In the rest of this section, we examine the important steps in the analysis process in more detail. The steps of Figure 5.5 we examine in detail are building a linearized circuit (line 4), determining the node voltage variables (line 5), ensuring that all element values can be computed

Constants
```
.PARAM cload = 1pF
.PARAM vddval=2.5
.PARAM vssval=-2.5
```

Topology
```
.subckt   oa        inpos     inneg     outpos    outneg    nvdd      nvss
M2        outpos    inneg     A         A         Ne        w='W'     l='L'
+         constraint1 = 'vgs - von > 0.2' constraint2 = 'vds - vdsat > 0.05'
M1        outneg    inpos     A         A         Ne        w='W'     l='L'
+         constraint1 = 'vgs - von > 0.2'constraint2 = 'vds - vdsat > 0.05'
R1        nvdd      outpos    2K
R2        nvdd      outneg    2K
ib        A         nvss      'I'
.ends
```

Variables
```
.VAR W RANGE=(1.8u,500u) GRID=1 RES=0.005 CLASSES=3 POSTNR=1 SANITY=32u
.VAR L RANGE=(1.2u,100u) GRID=1 RES=0.01 CLASSES=3 POSTNR=1 SANITY=3.2u
.VAR I RANGE=(1uA,1mA) RES=0.001 CLASSES=3 POSTNR=1 SANITY=10mA
```

OBLX Constants
```
.SYNTH example
```

Bias Circuit
```
.OblxCkt self_bias bias
xamp      inpos     inneg     outpos    outneg    nvdd      nvss      oa
vdd nvdd  0         vddval
vss nvss  0         vssval
rf1       inneg     outpos    1e6
rf2       inpos     outneg    1e6
.spec     SR        value 'I/(2 * (cload + xamp.m1.cdd))'good = 1Megbad = 10k
.endOblxCkt
```

AWE Circuit 1
```
.OblxCkt openloop awe
.bias self_bias
xamp      inpos     inneg     outpos    outneg    nvdd      nvss      oa
vdd       nvdd      0         vddval
vss       nvss      0         vssval
vag       inneg     0         0
vin1      inpos     inneg     0.000000e+00 ac 1
cl1       outpos    0 Cload
cl2       outneg    0 Cload
.pz tf V(outpos) vin1
.spec     ugf       value 'unity_gain_freq(tf)'good = 1Megbad = 10k
.obj      Adm       value 'dc_gain(tf)'      good = 1000bad = 10
.endOblxCkt
```

AWE Circuit 2
```
.OblxCkt commonmode awe
.bias self_bias
xamp      in        in        outpos    outneg    nvdd      nvss      oa
vdd       nvdd      0         vddval
vss       nvss      0         vssval
vin1      in        0         0.000000e+00 ac 1
cl1       outpos    0 Cload
cl2       outneg    0 Cload
.pz cm V(outpos) vin1
.spec     Acm       value 'db(dc_gain(tf))'   good = 0    bad = -10
.endOblxCkt

.END
```

FIGURE 5.4 Problem Description File for Circuit *Simple OTA* (line 6), determining equations for the current in each branch (line 7), writing device operating region constraints (line 8), and writing performance constraints (line 14). Note that many of these algorithms included here are not entirely new, but based upon standard circuit analysis techniques [Vlach 94] and programming language compilation techniques [Aho 87].

5.2.1 Building a Linearized Equivalent Circuit

The first step in Figure 5.5 that we explain is building a linearized circuit. Within the compiler, circuits are represented as graphs. For a circuit graph, $G = \{V, E\}$, the vertices (V) in the graph correspond in a one-to-one relationship with the nodes in the circuit and the edges (E) in the graph correspond to the branches in the circuit. Three kinds of graphs are manipulated during compilation, the graphs for the circuits as they are read from the description, graphs for linearized large-signal equivalent circuits, and graphs for linearized small-signal equivalent circuits. The large-signal equivalent circuits are used to determine the node voltages that will become independent variables (become part of $\underline{x}$) and write the KCL equations that determine the large-signal current error at each node. Small-signal linearized circuits are passed on to OBLX to use for the Newton-Raphson algorithm in operating point determination and to use by AWE to determine circuit performance.

1 Collect the user-defined variables
2 Save the annealing parameters and device models
3 for each bias deck
4   build the linearized large-signal equivalent circuit
5   determine the node-voltage variables
6   generate a string for each element value
7   generate a string that expresses the current in each circuit branch
8   generate a string for each device operating region constraint
9 for each AWE deck
10   link each active device to a device in a bias circuit
11   build the linearized small-signal equivalent circuit
12   generate a string for each element value
13 for each constraint / objective, i
14   generate a string for each $f_i(\underline{x}) / g_i(\underline{x})$ and $\hat{f}_i(\underline{x}) / \hat{g}_i(\underline{x})$.

FIGURE 5.5 Analysis of a Circuit Description

Both of the linearized graph types are derived from the original circuit graph using the fairly straightforward algorithm shown in Figure 5.6. Here, $\tilde{G} = \{\tilde{V}, \tilde{E}\}$ is the linearized graph. The key step is that the encapsulated device library is queried and small-signal and large-signal models are substituted for active devices to build the linearized circuit topologies. These models can be quite complex and include several internal nodes. For example, the model topologies returned by the BSIM MOSFET model are shown in Figure 5.7. Because these can be quite complicated, they can substantially increase the number of nodes and elements in the circuit. Since the topology is returned by the encapsulated device evaluator, the compiler is completely independent of the model used. No other synthesis system of which we are aware is as independent from the underlying device models.

5.2.2 Determining the Node-Voltage Variables

The next and possibly most important task performed by the compiler is to determine which additional variables need to be added to $\underline{x}$ to support the relaxed D.C. formulation. Recall that the annealer will directly manipulate node voltages, evolving towards a set of D.C. correct values as it evolves towards a circuit that meets the performance specifications. A set of linearly independent node voltages to include in $\underline{x}$ must be determined. To do this, we perform more extensive manipulations of the circuit graphs. In this case, we work with the linearized D.C. graph. The algorithm consists of the following steps:

1. Analyze the topology. Prepare for a tree-link analysis [Vlach 94] by determining which circuit elements belong in the tree and which will be links. The detailed algorithm is given in Figure 5.8.

for each edge $E_{ij} \in \tilde{E}$ incident on vertex i

    if E_{ij} corresponds to a linear element

        $\tilde{E} \leftarrow \tilde{E} \cup \{E_{ij}\}, \tilde{V} \leftarrow \tilde{V} \cup \{V_j\} \quad \forall V_j$ connected to $E_{ij}, V_i \neq V_j$

    else (it is a non-linear device)

        $\tilde{V} \leftarrow \tilde{V} \cup V_H, \tilde{E} \leftarrow \tilde{E} \cup E_H$, where $H = \{V_H, E_H\}$ is the
        subgraph returned by an encapsulated device evaluator

FIGURE 5.6 Algorithm to Linearize a Circuit

2. Create the circuit tree. The detailed algorithm is given in Figure 5.9.
3. Walk the path from each node to ground. Write an analytic expression for the node voltage at each node.

The basic strategy is to determine a unique path from each node to ground. This is accomplished with steps 1 and 2. An analytic expression for the voltage at each node can then be written by tracing along this path. If a voltage source is encountered, the voltage at one end of the source is clearly the voltage at the other end plus the value of the source. If a resistor or current source is encountered along the path, the node voltage is not as easily determined. In this case, we choose to add the voltage to the set of independent variables. In early versions of the compiler we performed as many simplifications as possible, using Ohm's law wherever possible to make the voltage at one end of a resistor dependent on the voltage at the other end. However, we discovered that the Newton-Raphson algorithm worked best when the complete basis set of

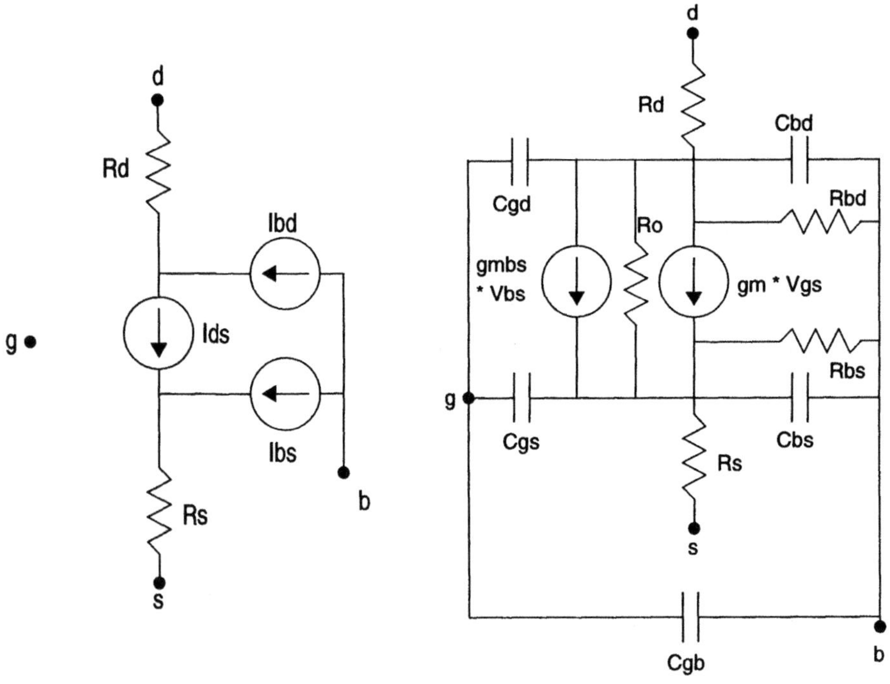

FIGURE 5.7 D. C. Large-Signal (left) and Small-signal (right) Topologies Returned by BSIM MOSFET Model

node voltages was manipulated. This is because KCL error information is calculated only at nodes with independent voltage variables. We found that the algorithm worked better with more complete gradient information. As a result, we now add a new variable whenever anything but a voltage source is encountered while tracing the path.

Start with circuit graph, V vertices (nodes), E edges (elements)
Sort the elements by type, voltage sources first, current sources last
Mark all elements as not in the tree
Initialize Γ, a set of sets of nodes s.t. $\Gamma_i = \{V_i\}$.
for each edge, E_i
    Examine its 2 vertices, V_+ and V_-
    if V_+ and V_- are in different elements of Γ
        $\Gamma_+ = \Gamma_+ + \{V_-\}$
        mark E_i as being in the tree

FIGURE 5.8 Algorithm for Topological Analysis

Start with results of Topological Analysis (Figure 5.8)
Mark all elements as not seen
Initialize a Stack of Nodes, S, with the ground node
while S is not empty
    node = Pop S
    for each edge, E_i, incident on *node*
        if (E_i not seen and E_i in tree)
            Mark E_i as seen
            Mark E_i as the path from *node* to ground
            Push the vertex at the other end of E_i on to S

FIGURE 5.9 Algorithm for Making a Circuit Tree

5.2.3 Ensuring That All Element Values Can Be Computed

Recall that an expression can be given for any numeric value in the input description. Each of these expressions must be eventually replaced by a number that is updated for each circuit evaluated by the annealer. For circuit element values the number eventually acts as input to AWE. For device parameters, the number becomes input to an encapsulated device evaluator. In general, any of these expressions may include references to user-defined variables, as a result the number will change each time the variables change value. In this case, the compiler must generate C code to re-evaluate the number each time the variable changes. However, in many cases, the expression can be resolved completely by the compiler. Here, the compiler interprets the expression directly and passes a constant on to the annealer.

The techniques we use to process each expression are quite similar to standard compiler algorithms [Aho 87]. First we parse the expression into tree form, using a standard recursive descent parser. We then walk the tree and generate a string of C code that will evaluate it. Where possible, we interpret the expression. If the entire expression can be interpreted, the generated code is a single constant. If a variable or function is encountered that can be evaluated only at run-time, the string will include appropriate references to live variables or executable functions.

Except for function calls and variable references, generating the string is done with a quite straightforward in-order traversal of the tree [Cormen 90]. Function names and variable names are maintained in a hash table for ready access. When a function is encountered, the string generation is called recursively on the function's arguments. When a variable is encountered, it must be checked to see if it is a reference to a device parameter (*e.g.*, `xamp.m1.vdsat` from Figure 5.4). In this case, we must expand the name and perform a recursive search through the circuits and sub-circuits defined in the input trying to find a device that matches. After finding a device with the matching name, we then verify that the parameter (*e.g.*, vdsat) is available from the encapsulated device evaluator. Finally, we generate C code that accesses the appropriate parameter of the device each time a new move is evaluated.

5.2.4 Determining Equations for the Current in Each Branch

Recall from Section 4.4.5 that the relaxed D.C. formulation, represented by the cost terms in C^{dc}, requires the KCL error currents at each node. Given a node k and its set of incident branches ζ^k, we calculate the KCL error current as:

$$\Delta i_k = \sum_{b \in \zeta^k} I_b \tag{5.4}$$

$$mag_k = \frac{1}{\|\zeta^k\|} \cdot \sum_{b \in \zeta^k} |I_b| \tag{5.5}$$

The cost terms in C^{dc} can be computed from these equations as discussed in Section 4.4.5.

We have already generated analytic expressions for the voltage at each node (Section 5.2.2) and for each element value (Section 5.2.3). Given these expressions, we can write expressions for the current through each branch. For current sources the current flow is simply the element value and for resistances Ohm's law is used. Voltage sources are typically not encountered because the error current is calculated only for nodes whose voltages are independent variables, since there is no Kirchhoff's Law error if this voltage is dependent on another voltage. Since independent node voltage variables were created only when the path to ground did not go through an independent voltage source, encountering voltage sources in the branch current calculation occurs only with dependent sources. However, we do handle this case by using the sum of currents leaving the elements attached to the other end of the source.

5.2.5 Writing Device Operating Region Constraints

The next algorithm in the circuit compiler that warrants additional explanation is that used to write device operating region constraints. Recall that each active device may include two relational expressions that specify operating region constraints. For example, from line 9 of Figure 5.4, the expression 'vds - vdsat > 0.05'. Recall from Section 4.4.4 that expressions of this form are rearranged into the following form:

$$g_i(\underline{x}) \leq 0 \tag{5.6}$$

This is done by manipulating the expression tree before it is interpreted. The modified expression tree is then processed as discussed in Section 5.2.3 above.

5.2.6 Generating C Strings to Evaluate the Specifications

The final aspect of the analysis phase is to examine all the functions that specify constraints and objectives, generating C strings that evaluate the specifications value. Again, this process is much as the C string generation discussed in Section 5.2.3 above.

5.3 Code Generation

As the last step in compilation, all the code that is both specific to the circuit being synthesized and necessary to evaluate the cost function must be generated. Whenever possible, the OBLX library has been designed to operate in a circuit-independent fashion, to minimize the amount of code that must be generated and compiled. Nevertheless a substantial amount of output is generated. We can categorize the code as being for either *initialization*, *cost function evaluation*, or *output generation*. Refining these categories further, the initialization requires:

- Small-signal circuit topologies to initialize AWE.
- Large-signal circuit topologies to initialize the Newton-Raphson Algorithm.
- Code to initialize the annealer's state.
- Code to initialize the independent variables.

Cost function evaluation requires this circuit-specific code:

- Code to pass device parameters and node voltages to the encapsulated device evaluators.
- Code to update any linear elements that depend on $\underline{x}$ in AWE.
- Code to evaluate the KCL error at each node.
- Code to evaluate the device region constraints.
- Code to evaluate the user-specified constraints and objectives.

Output generation requires this circuit-specific code:

- Code to generate a circuit simulation deck than can be used to verify the circuit performance.

5.4 Implementation

The compiler we have discussed in this chapter has been implemented as a program called ASTRX. The code it generates interfaces directly to OBLX, AWE, and the encapsulated device evaluators. Together, ASTRX/OBLX provide a complete path from input topology and performance specifications to a sized circuit schematic, implementing the flow shown in Figure 3.2 on page 63.

As with OBLX, ASTRX is written in a highly structured fashion. The main body of code is built upon several layers of software libraries. The overall software structure is shown in Figure 5.10. Descriptions of the various components of the overall structure are given in Table 5.3 .

5.5 Summary

In this chapter we discussed the details of the circuit compilation process and its implementation as ASTRX. ASTRX reads a single description file that follows a SPICE-like format. This file contains all the information needed to generate the problem-specific cost function evaluation code that is linked to OBLX to create a synthesis tool customized to solve the circuit synthesis problem detailed by the user. Included in this description file is a list of user-specified variables, a set of performance specifications, and sets of circuits that specify bias and AWE analysis conditions for the circuit under design. If a numeric value in these descriptions is replaced with an expression that depends upon one or more user-specified variables, ASTRX generates code that re-evaluates that value each time the annealer generates a move. To implement the relaxed D.C. formulation, ASTRX also analyzes the bias circuits

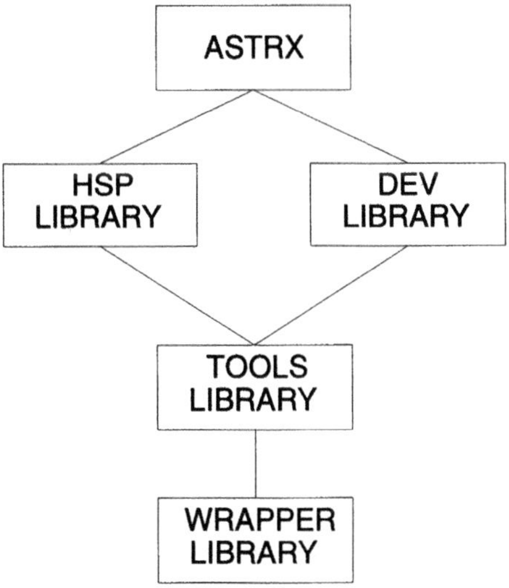

FIGURE 5.10 Software Structure of ASTRX

and generates a basis set of independent node voltages that are then added to the set of independent variables manipulated by OBLX. Finally, the compiler generates all the code necessary for the annealer to manipulate a basis set of bias voltages, invoke device models to evaluate new bias currents and small-signal models, invoke AWE to analyze the overall behavior of the circuit, and evaluate the complex multiple-component cost function that defines the quality of each circuit configuration visited. This code is then compiled and linked with the OBLX library to form an executable program that designs the circuit to meet the user-specifications.

Now that we have completed our description of the nominal synthesis formulation used in ASTRX/OBLX, in the next chapter, we present synthesis results for a variety of circuits designed automatically with ASTRX/OBLX.

TABLE 5.3 Components of ASTRX

Name	Description	Origin	Source Lines
ASTRX	Main compilation routines for ASTRX. All analysis routines and code generation.	for ASTRX	12700
HSP	A reconfigurable parser for SPICE-like languages.	for ASTRX	12700
DEV	Non-linear circuit device models adapted from SPICE. Presently includes BSIM and MOS3 models for MOSFETs and Gummel-Poon model for BJTs.	SPICE3/ for OBLX	26200/ 3000
Tools	Programmer's Tool-kit. Portions used include: dynamic arrays, hash tables, lexical analyzer, execution manager, error manager, and initialization file manager.	for ASTRX/ OBLX	13300
Wrapper	A portability layer built on the UNIX library.	for ASTRX/ OBLX	3900
	Total / Total specifically for ASTRX/OBLX		71800/ 45600

CHAPTER 6

Nominal Circuit Synthesis Results

To validate our nominal synthesis formulation, we organize nominal synthesis results into two chapters that correspond to the broad directions in which we have tested its implementation in ASTRX/OBLX. First, in this chapter, we present nominal circuit synthesis results to show that ASTRX/OBLX can synthesize analog circuits and that—when compared to prior approaches—it has improved accuracy, automation, and breadth. The second class of results, presented in Chapter 7, show that the engineering decisions made during the design and implementation of ASTRX/OBLX are effective. In that later chapter we include results that show the efficacy of simulated annealing in our application, our dynamic weighting mechanism, the move control mechanisms and the relaxed-D.C. formulation. In this chapter we organize our presentation as follows. First we focus on comparing the automation and accuracy of ASTRX/OBLX to prior approaches. For a suite of circuits, we compare the synthesis results from other tools with results generated by ASTRX/OBLX. Second, we focus on generality/scalability by showing the breadth of circuit synthesis tasks that ASTRX/OBLX can perform. To show the utility of the encapsulated evaluators, we present designs for different processes, different models, and different non-linear devices. To show the ability of ASTRX/OBLX to quantify the performance potential of a circuit topology, we show how it can be used to explore the space of achievable performance targets for a given circuit topology. To show that ASTRX/OBLX circuit designs compare favorably with full-custom manually designed circuits, we compare a highly optimized manual circuit design to a re-synthesis of the same circuit with

ASTRX/OBLX. And finally, to show that ASTRX/OBLX scales well and is capable of designing large circuits, we present the largest analog cell synthesis results published to date.

6.1 Previously Published Synthesis Results

Table 6.1 lists the synthesis tools discussed in Section 2.4 and the circuits for which they have published synthesis results. These five circuits comprise a comparison suite because they represent the bulk of all synthesis results ever published. In this section, we use ASTRX/OBLX to re-synthesize *all* these circuits, thereby duplicating the bulk of the previously published synthesis results. Note that in some published papers circuit types are merely mentioned and schematics and results are not provided. In other cases, the circuit schematics published are minor variations on the schematics that appear here. In both cases, we make the reasonable simplification that the circuits are equivalent to the similar circuits described here. That few published results exist and that they are limited in scope and performance is perhaps one of the best indicators that the first-time effort to design a circuit has always been a substantial barrier to overcome. Researchers have typically published synthesis results for fairly well understood circuit topologies, presumably because of the ready availability of circuit performance equations.

TABLE 6.1 Prior Approaches and the circuit examples published

	Circuits Published				
Synthesis Tool	**Simple OTA**	**OTA**	**Two-Stage**	**Folded Cascode**	**Comparator**
OPASYN			■	■	
OASYS			■		■
IDAC[a]		■			■
ARIADNE			■	■	
STAIC	■			■	
ISAID			■		
Maulik/Carley			■		

a. IDAC can design several other circuits not included in this table.

6.1.1 Circuit Descriptions

Before we present synthesis results in detail, we first describe the circuits in our test suite. For all of the results that follow, some manually derived equations were included in the ASTRX input deck to model large-signal specifications that cannot be readily or efficiently evaluated using our implementation of AWE. In these cases, the equations we included are noted in a table near the circuit schematic. Note that these equations themselves and the circuit analyses are not the primary contribution of our work but are necessary to fully specify realistic synthesis examples. As a result, our goal when deriving performance equations was to determine a reasonably accurate equation as quickly as possible. It is also important to note that using AWE allows the designer to complete equation derivation quickly because the bulk of the equation derivation task is avoided. This is because the difficult equations are those that describe small-signal behavior of the circuit. For equation-based tools, these equations must be derived and coded into the tool, but for ASTRX/OBLX small-signal performance is determined via simulation with AWE.

Simple OTA

We refer to the first circuit in the comparison suite as *Simple OTA*. It is a very simple operational transconductance amplifier, and as such an easy first synthesis target. The schematic diagram is shown in Figure 6.1.

The manually derived equations added to the input description are detailed in Table 6.2. To complete the ten specifications of interest to the designer, three equations must be derived. The first of these, slew rate, is determined from inspection at the output node. With a large differential input, essentially all the bias current generated in M5 will be used to drive the capacitance at the output node. The slew rate is then calculated as simply *I/C*. Note that this expression relies heavily on the assumption that *I* remains constant. This assumption is valid only if the transistors providing the current remain in saturation during slewing. We can force the annealer to design this characteristic into the circuit by adding additional saturation constraints to the problem description. When actually writing the expression, both the current in M5 and each of the contributing capacitances at the output node are available as predefined variables in the input deck, making the slew rate equation in Table 6.2 straightforward to write. To determine the output swing, two equations are required, one for the swing up and the other for the swing down. In each case, a heuristic is used. It is assumed that the drain-source voltage on the device can shrink until it reaches Vd_{sat}, the voltage required to keep the device in saturation. In practice these equations are not exact and can lead to differences of about 10% between predicted and simulated performance. However, it is important to restate that these equations are quite simple to derive and yet yield acceptable results; this would not be the case for small-signal specifications such as gain and phase margin.

OTA

The second circuit in our suite is the more common variation of an operational transconductance amplifier that we shall refer to simply as circuit *OTA*. Figure 6.2 is the circuit schematic for the circuit as synthesized by ASTRX/OBLX.

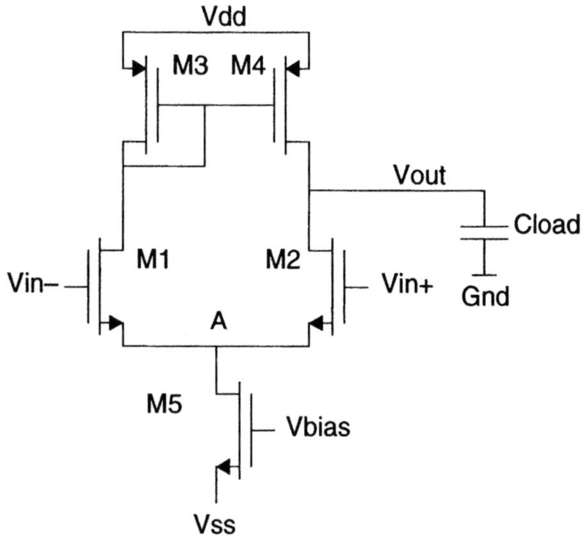

FIGURE 6.1 Circuit *Simple OTA*: Schematic Diagram

TABLE 6.2 Circuit *Simple OTA*: Performance Equations Added.

Specification	Equation
Slew Rate	$\frac{M5.Id}{Cload + M4.Cd + M2.Cd}$
Output Swing (Up)	$M4.Vds - M4.Vdsat$
Output Swing (Down)	$M2.Vds - M2.Vdsat$

As with *simple OTA* we add three performance equations, which are detailed in Table 6.3. Again, the slew rate is calculated as the quotient of the tail current and the capacitance at the output node. Also, as before, the output swing is computed by taking the difference between the nominal drain-source voltage and Vd_{sat} for the output devices.

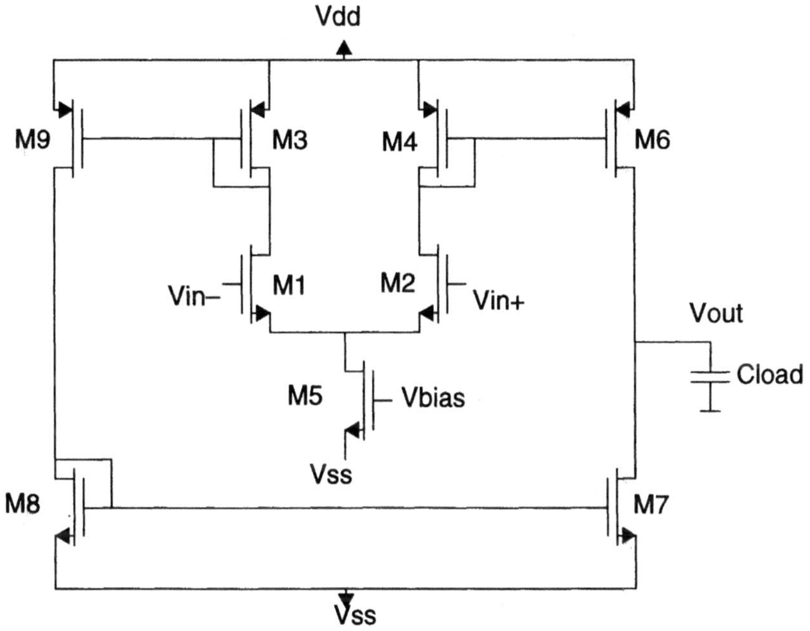

FIGURE 6.2 Circuit *OTA*: Schematic Diagram

TABLE 6.3 Circuit *OTA*: Performance Equations Added

Specification	Equation
Slew Rate	$\frac{M5.Id}{Cload + M6.Cd + M7.Cd}$
Output Swing (Up)	$M6.Vds - M6.Vdsat$
Output Swing (Down)	$M7.Vds - M7.Vdsat$

Two-Stage

Figure 6.3 is the circuit schematic for the two-stage operational amplifier, we shall call *Two-Stage*, as synthesized by ASTRX/OBLX. The manually derived equations added to the input description are detailed in Table 6.4. The procedure followed for their derivation is as described for the two other opamps. However, when determining output swing, one additional assumption, that only the drain-source voltage across the two cascode devices will vary, must also be made.

Folded Cascode

The fourth circuit in the suite is again an operational amplifier, but in this case it is a fully differential amplifier and includes common-mode feedback and bias circuits. We shall call this circuit *Folded Cascode*. The schematic diagram is shown in Figure 6.4.

The manually derived equations added to the input description are detailed in Table 6.5. The derivations are similar to those for the other amplifiers.

Comparator

Figure 6.5 is the circuit schematic for the comparator synthesized by ASTRX/OBLX. This circuit is quite important to the suite because it is the only circuit that is not an operational amplifier. Note that with the exception of OASYS and IDAC, existing synthesis tools design only operational amplifiers. The performance of a comparator is much more difficult to characterize. This is also the case within ASTRX/OBLX because the circuit has important performance characteristics that are not readily characterized by the small-signal transfer functions automatically generated by AWE.

The primary design concern in this comparator is the time it takes for the latches (the cross-couple MOS devices controlled by $\phi 1$ and $\phi 3$) to set. When the first stage latch sets, one of the nodes A and B is driven to ground while the other rises to Vdd. For the node that swings down, the settling time can be determined with AWE. However, this is a particularly challenging use of AWE because the latch employs positive feedback and so will have a positive pole in its transfer function. As discussed in Section 4.4.7, multiple AWE approximations can be performed for a single circuit to ensure that a numerically unstable Padé approximation does not corrupt circuit performance predictions. For circuits that cannot have positive poles, unstable Padé approximations typically manifest as approximations with positive poles. For these circuits, we employ an aggressive "paradiagonal" strategy (as described in Section 4.4.7) that performs a small number of approximations where each approxi-

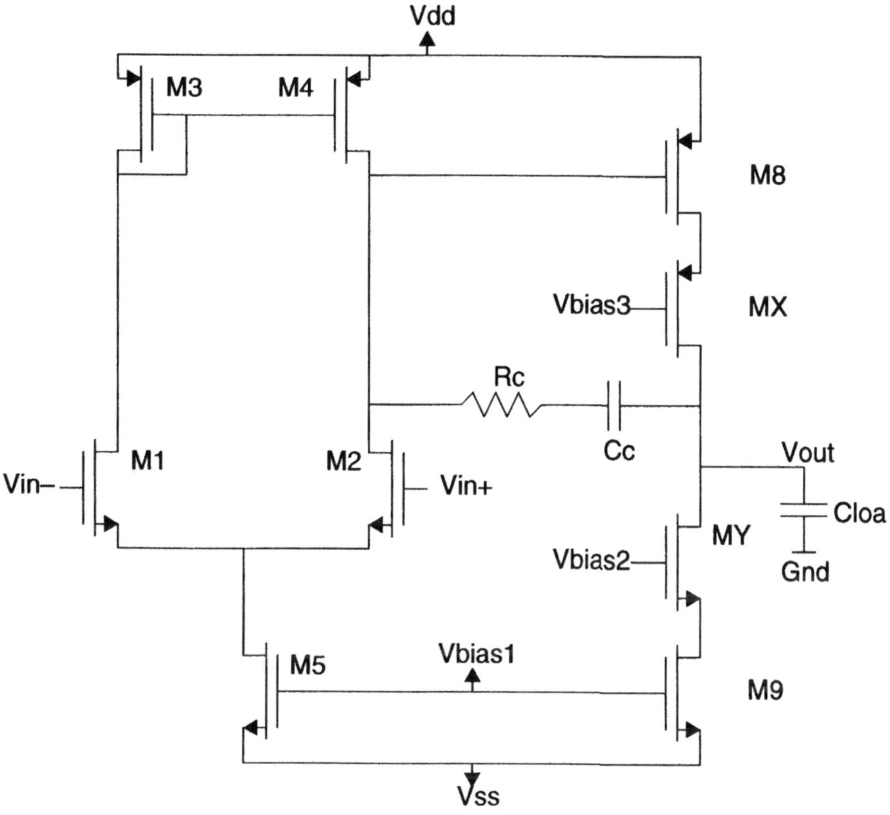

FIGURE 6.3 Circuit *Two-Stage*: Schematic Diagram

TABLE 6.4 Circuit *Two-Stage*: Additional Equations Added

Specification	Equation
Slew Rate	$\frac{MY.Id}{Cload + Cc + MY.Cd + M\dot{X}.Cd}$
Output Swing (Up)	$MX.Vds - MX.Vdsat$
Output Swing (Down)	$MY.Vds - MY.Vdsat$

mation attempts to find one less zero than pole. Approximations with positive poles are simply skipped. For the comparator, ASTRX/OBLX switches to a more time-consuming "horizontal" search strategy (see Section 4.4.7) that attempts to guarantee a pole's location by performing multiple approximations with the same number of poles but differing numbers of zeros. Using this technique, we can be confident that positive poles found in each of a number of different approximations are actually present in the circuit. Thus, even for this circuit with positive feedback, we can employ AWE to automate the performance evaluation.

However, the behavior of the node that swings to Vdd when the latch closes is large-signal and cannot be modelled with AWE. This information is captured by the slew rate at this node, which we derive in much the same fashion as for the amplifier cir-

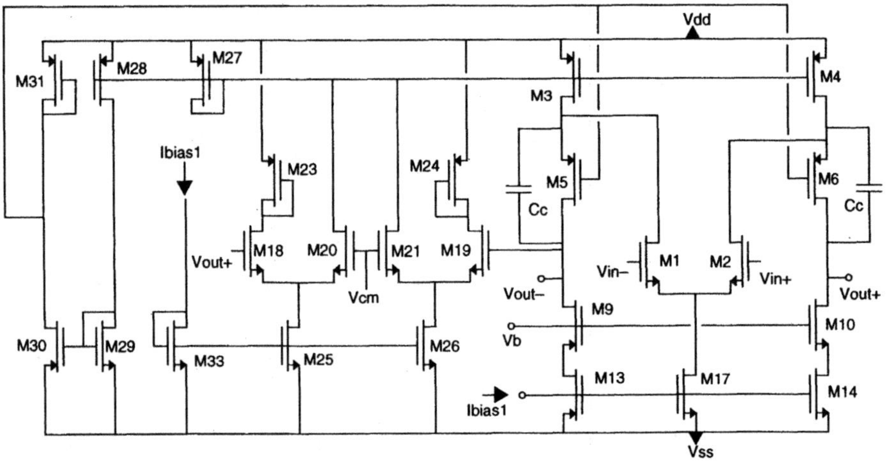

FIGURE 6.4 Circuit *Folded Cascode*: Schematic Diagram

TABLE 6.5 Circuit *Folded Cascode*: Additional Equations Added

Specification	**Equation**
Slew Rate	$\frac{M4.Id}{Cload + Cc + M6.Cd + M10.Cd}$
Output Swing (Up)	$M6.Vds - M6.Vdsat$
Output Swing (Down)	$M10.Vds - M10.Vdsat$

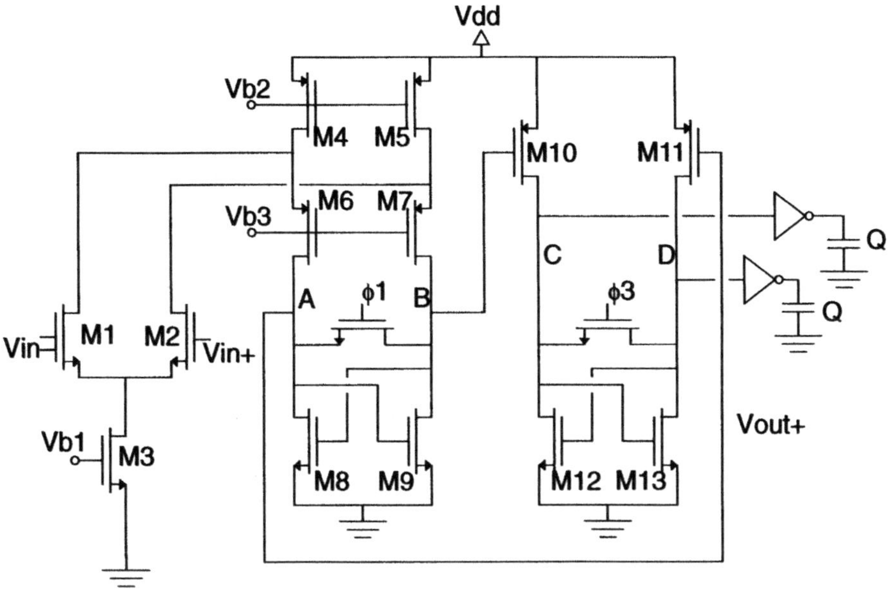

FIGURE 6.5 Circuit *Comparator*: Schematic Diagram

TABLE 6.6 Circuit *Comparator*: Additional Equations Added

Specification	**Equation**
Slew Rate (A)	$\frac{M8.Id}{M8.Cd + M\phi1.Cs + M6.Cd + M10.Cg}$
Slew Time (A)	$\frac{Vdd - V(A)}{\text{Slew Rate(A)}}$
Slew Rate (C)	$\frac{M10.Id}{Cbuf + M\phi3.Cs + M12.Cd + M10.Cd}$
Slew Time (C)	$\frac{Vdd - V(C)}{\text{Slew Rate(C)}}$
Error Rate	$KA_1A_2e^{\frac{Tp1}{2}}e^{\frac{Tp2}{2}}$

cuits. The manually derived equations added to the input description are detailed in Table 6.6. Because we are actually interested in the time it takes for the node to slew to Vdd, we also include a Slew Time specification as shown in the table. The final equation in the table is the error rate for the circuit. This is captured by the third equation in Table 6.6, which gives the error rate in mean seconds between errors. Here, the constant, K, is the ratio of the designed common mode input range to the input range required to switch one of the two buffers that act as the comparators output stage. The gains A_1 and A_2 are the gains of the first and second stage and are computed automatically with AWE. Similarly, $p1$ and $p2$ are the positive poles from the latches of the first and second stages which were returned by AWE, while T is the clock period specified for the circuit's operation.

ASTRX's Analyses

Table 6.7 summarizes the circuits in our comparison suite and lists the results of ASTRX's analysis of each. The number of lines in the input description is the sum of the first two rows. We separate this into two values. The first value is the number of lines required to describe the netlist of the circuit. This is essentially all the information that would appear in a simulation deck for the circuit and all the test configura-

TABLE 6.7 Result of ASTRX's Analyses for Comparison Suite

		Circuit				
		Simple OTA	OTA	Two-Stage	Folded Cascode	Comparator
Input (lines)	Netlist/Models	30	34	43	65	131
	Synth. Specific	28	33	40	56	68
x	User-Supplied	7	11	19	28	19
	Node Voltages	14	24	26	70	57
$C(x)$	Terms	56	85	88	212	169
	Lines of C	1443	1809	1894	3408	3088
Circuits	Circuit Type[a]: nodes, elements	B: 20, 31	B: 28, 49	B: 34, 54	B: 75, 138	B: 65, 126
		A: 20, 67	A: 29, 114	A: 33, 118	A: 75, 324	A: 63, 265
					A: 75, 324	A: 64, 266
						A: 29, 115

a. Type 'A' is a linearized AWE circuit. Type 'B' is a bias circuit.

tions used to measure performance, including the circuit elements and device models. The second value is the number of lines specific to ASTRX/OBLX. This includes specifying the independent variables and performance specifications. For both numbers, comments were not counted and logical lines that were spread over multiple physical lines of the file by using the SPICE continuation character were counted only once. In all, the amount of time and effort required to create an ASTRX/OBLX input description is quite modest. It is typically more difficult to ensure that the topology is correct than it is to provide the additional synthesis specific information required by ASTRX/OBLX.

From this input description, ASTRX generates the cost function, $C(\underline{x})$, as C code that will be compiled and linked with the OBLX library. The rest of Table 6.7 summarizes the information generated by ASTRX: optimization variables, cost function specifics and linearized circuit sizes. The sum of the third and fourth rows is the total number of independent variables, the number of elements in $\underline{x}$, that are manipulated by OBLX when minimizing $C(\underline{x})$. Row three is the number of user-specified variables, which are simply read from the input deck. The fourth row is the number of independent variables added by ASTRX to codify the relaxed D.C. formulation. Recall that each linearly independent node voltage becomes an independent variable. Note that because the topologies of the device models we employ contain internal nodes, these node voltage variables typically outnumber the user-specified variables, increasing the dimensionality of the optimization problem. However, this is offset somewhat by the inclusion of directed Newton-Raphson moves in OBLX's move-set, since these are gradient-based and move the circuit towards D.C. correctness. In some sense, the Newton-Raphson moves make it easier to find values for the node voltages than it is to find values for the user-specified variables. Row 5 shows the number of terms in the cost function generated by ASTRX. This includes a term for each performance specification, each node variable, and each device region constraint. In total, the number of lines of C generated by ASTRX is shown in row 6. This is the sum of the lengths of three files generated. Although all the code common to every circuit is stored in the OBLX library, as described in Chapter 5, ASTRX must still generate a substantial amount of C. Finally, the last rows in Table 6.7 show the size of the linearized small single circuits generated by ASTRX. Recall that bias circuits are used to determine D.C. operating points while AWE circuits are passed to AWE for small-signal performance evaluation. Recall also that ASTRX generates these circuits by linearizing the bias and AWE circuits provided by the user in the input description. The number of nodes in these circuits is crucial to the overall run-time of OBLX. The matrices factored by AWE and used during Newton-Raphson gradient voltage perturbation are proportional in size to the number of nodes, and as a circuit gets large the run-time of these algorithms is dominated by this factorization, which ranges from

$O(n^{1.4})$ to $O(n^{1.7})$ [Raghavan 93] for an n-by-n matrix, depending on the circuit and the implementation.

6.1.2 Synthesis Results

Basic circuit synthesis results for our comparison suite are contained in Table 6.8, Table 6.9, and Table 6.10. Each result shown here is the best of several runs. Because of the stochastic nature of annealing, the most practical procedure to follow when designing a circuit is to perform five to ten optimization runs and select the one(s) that best meet the designer's needs. For this reason run-times given should be multiplied by ten to obtain a fair comparison with other synthesis systems. Run-times reported are for an IBM RS/6000 550.

TABLE 6.8 Basic Synthesis Results for Comparison Circuits, BSIM Models, 1.2μ Process

	Specification: OBLX / Simulation					
Attribute	**Simple OTA**		**OTA**		**Two-Stage**	
Cload (pF)		1		1		1
Vdd (V)		5		5		5
D.C. gain (dB)	↑[a]:	38.1 / 38.2	↑:	40.4 / 40.2	≥60:	66.4 / 66.4
gain bandwidth (MHz)	≥50:	50.1 / 50.1	≥25:	25.0 / 25.4	≥10:	10.6 / 10.6
phase margin (°)	≥60:	76.6 / 75.6	≥45:	57.9 / 57.8	≥45:	87.3 / 86.5
PSRR (Vss)	≥20:	68.7 / 70.3	≥40:	42.1 / 42.0	≥20:	31.0 / 30.9
PSRR (Vdd)	≥20:	38.5 / 38.7	≥40:	52.8 / 52.8	≥40:	45.8 / 45.8
output swing (V)	≥±1.0:	±1.4 / ±1.2	≥2.5:	4.0 / 4.0	≥2:	2.7 / 2.8
slew rate (V/μs)	≥50:	92.5 / 89.4	≥10:	51.6 /48.2	≥2:	3.8 / 4.0
active area ($10^3 \mu^2$)	↓:	1.2	↓:	0.9	↓:	2.1
static power (mW)	≤1:	0.51 / 0.51	≤1:	0.33 / 0.34	≤1:	0.16 / 0.16
time/ckt. eval (ms)		41		37		38
CPU time (min. / run)		5		9		16

a. ↑ means maximize, while ↓ means minimize.

6.1.3 Comparing Automation and Accuracy

Direct comparison of the performance of ASTRX/OBLX synthesized circuits and those of previous synthesis tools is difficult because of the unavailability of device model parameters to describe the processes for which they were designed.As discussed in Section 1.4, this limitation can be addressed by using accuracy as an indirect measure of performance. To provide a quantitative measure of accuracy, we compare the performance of the synthesized designs as predicted by OBLX with the results of simulation. Since published results for prior approaches usually make similar comparisons, we can use the difference from simulation as an error metric that measures the tool's accuracy. In this section we compare prior approaches with ASTRX/OBLX, on the basis of accuracy and two measures of automation—preparatory effort and synthesis time. Table 6.11, Table 6.12, Table 6.13, Table 6.14, and Table 6.15 provide a comparison of first-time designer effort, run-time and worst-case error between other tools and ASTRX/OBLX. The preparatory effort required to

TABLE 6.9 Circuit Folded Cascode: Synthesized Result, BSIM Models, 2μ Process

Attribute	Specification:	OBLX/Simulation
Cload		1.25pF
Vdd		5V
D.C. gain (dB)	≥70:	70.1 / 70.1
gain bandwidth (MHz)	maximize:	72.4 / 72.1
phase margin (°)	≥60:	80.0 / 80.0
PSRR (Vss, dB)	≥105:	107 / 107
PSRR (Vdd, dB)	≥105:	125 / 125
CMRR (dB)	≥92:	97 / 97
output swing (V)	≥±1.0:	±1.5 / ±1.5
slew rate (V/μs)	≥50:	67 / 57
active area ($10^3\mu^2$)	minimize:	46
static power (mW)	≤15:	10 / 10
time/ckt eval (ms)		116
CPU time (min. / run)		120

TABLE 6.10 Circuit Comparator: Synthesized Result, BSIM Models, 1.2μ

Attribute	Specification:	OBLX/Simulation
Cload		1.25pF
Vdd		5V
stage 1 settling, 0.1%, (ns)	≤5:	4.5 / 4.7
stage 2 settling, 0.1%, (ns)	≤5:	4.6 / 4.8
stage 1 slew rate (V/μs)	≥800:	800 / 750
stage 2 slew rate (V/μs)	≥800:	800 / 620
latch 1, positive pole (MHz)	≥200:	430 / 450
latch 2, positive pole (MHz)	≥200:	360 / 330
error rate, (1 in years)	≤10:	8E+24
active area ($10^3\mu^2$)	minimize:	1.6
static power (mW)	minimize:	2.7 / 2.7
time/ckt eval (ms)		136
CPU time (min. / run)		97

TABLE 6.11 Circuit Simple OTA: Automation

Synthesis Tool	Characteristic: Preparatory Effort	CPU Time	Worst Case Err. vs. Sim.
STAIC	This circuit and 2 variations of folded cascode took 10,000 lines of code to describe.	3 min. MIPS 2000	24%
ASTRX/ OBLX	1 hour 58 lines of input	50 min. / 10 runs IBM RS/6000	13% (eqns) 2% AWE specs

TABLE 6.12 Circuit OTA: Automation

Synthesis Tool	Characteristic Preparatory Effort	CPU Time	Worst Case Err. vs. Sim.
IDAC	9 man-months. About 250 lines of FORTRAN.	"A few seconds"	15%
ASTRX/OBLX	2 hours 58 lines of input	90 min. / 10 runs IBM RS/6000	16% (eqns) 2% AWE specs

TABLE 6.13 Circuit Two Stage: Automation

Synthesis Tool	Characteristic Preparatory Effort	CPU Time	Worst Case Err. vs. Sim.
OPASYN	A few days to add the equations, plus an unspecified amount of time to tune the "fitting functions"	< 5 min. / run DEC 8800	200% un-tuned 20% tuned
OASYS	6 - 9 months 4000 lines of C	< 10 sec. DEC 3100	7%
ARIADNE	46 design equations, some derived with the aid of ISAAC	5 min. platform N/A	7%
ISAID	N/A	N/A	14%
Maulik/Carley	6 months 82 constraint equations	1 min. DEC 3100	16%
ASTRX/OBLX	3 hours 83 lines of input	3 hrs. / 10 runs IBM RS/6000	5% (eqns) 1% AWE specs

TABLE 6.14 Circuit Folded Cascode: Automation

Synthesis Tool	Characteristic Preparatory Effort	CPU Time	Worst Case Err. vs. Sim.
OPASYN	2 weeks to add the equations, plus an unspecified amount of time for tuning the "fitting equations"	5 min. DEC 8800	200% un-tuned 20% tuned
STAIC	2 variations of this circuit and a Simple OTA took 10,000 lines of code to describe.	11 min. SUN Sparc 2	20%
ARIADNE	N/A. Derivation is manual but aided by symbolic simulator ISAAC	1 min. VAX 750	15%
ASTRX/OBLX	1 day 121 lines of input	20 hrs. / 10 runs IBM RS/6000	17% (eqns) 0.4% AWE specs

TABLE 6.15 Circuit Comparator: Automation

Synthesis Tool	Characteristic Preparatory Effort	CPU Time	Worst Case Err. vs. Sim.
OASYS	9-12 months 1500 lines of lisp	5 sec. DEC 8800	N/A
IDAC	12 months, about 45 lines of FORTRAN per device	N/A	N/A
ASTRX/OBLX	2 days 199 lines of input	19 hrs. / 10 runs IBM RS/6000	22% (eqns) 9% AWE specs

create a typical circuit using these previous approaches is substantial, varying from weeks to years, depending upon how well the circuit is understood before the process begins. However, after this initial design phase, most of the tools run within a few minutes. In contrast, with ASTRX/OBLX, a typical new design takes an afternoon of preparation followed by a set of optimization runs that can take overnight. We feel that the ASTRX/OBLX scenario is more appealing to the average designer: after an afternoon of effort, the designer goes home while the synthesis tool completes the work.

Because of the simplifications made during the circuit analysis, results from equation-based synthesis tools typically differ from simulation by between 10% and 200%. In contrast, because of AWE's inherent accuracy, ASTRX/OBLX results typically match simulation results exactly. In the tables, ASTRX/OBLX worst-case error when compared to simulation is given as two separate numbers. The first—and larger—number compares all specifications, including those for which simple 1st order equations were used. The second number compares only specifications for which AWE was used to predict performance. For specifications where AWE is used to predict performance, only rarely do predicted values differ from simulation by more than 2%. When manually derived equations are used, the accuracy is not as good and errors may be as large as 20%. When comparing AWE/OBLX predictions to simulation, it is important to note that our simulation results were obtained using HSPICE [Meta 87] while our models were adapted from Berkeley SPICE. Model differences are almost always the source of discrepancy between OBLX predictions and simulation. Although HSPICE also began with the Berkeley device models, they have modified them slightly to improve convergence and fix other minor errors. When comparing performance predications for linearized circuits (after the non-linear devices have been replaced with linear small-signal models) we find that AWE results match HSPICE results exactly. Although another possible source of error is the inherent difference between the reduced-order approximation generated by AWE and the actual response, in practice, a few poles are enough to predict response extremely accurately, and errors of this kind are infrequent. For identical models and bias points AWE/OBLX and SPICE-like simulation methods typically produce identical performance predictions.

For the simple equations we added to handle large-signal performance characteristics, the error is on par with the errors seen in the performance predictions published for prior approaches. This is as expected since equations are used in both cases. Thus, real analog design expertise is required to take full advantage of the capabilities of ASTRX/OBLX. We could reduce the discrepancy between our simple equations and simulation by adding second order terms to improve the accuracy of equations. However, to show the utility of the ASTRX/OBLX system, the additional effort required to improve these equations is not warranted.

As a final note on these accuracy comparisons, we make the following qualitative observation contrasting our design goals and those of previous approaches. Typically, we tend to aim for higher nominal performance than most published results. Some previously published results seem to be "easy" designs that did not really approach the limits of the circuit topology within a reasonable fabrication process. The simulated performance of these designs typically matches simple equation-based performance because the approximations made during equation derivation tend to remain valid. In contrast, higher-performance designs and more aggressive fabrication processes with smaller feature sizes tend to invalidate many equation simplification techniques. As a result, accuracy comparisons between low-performance circuits designed with equation-based tools and high-performance circuits designed with ASTRX/OBLX may favor the equation-based tools.

6.2 Device Encapsulation

Our second set of results focuses on the encapsulated device evaluators. To show the utility of encapsulated device evaluators, our library presently contains 3 different device models adopted from version 3e of SPICE [Johnson 91]: the level 3 MOS model, and the BSIM MOS model [Sheu 87] and the Gummel-Poon model for BJT devices [Gummel 70, Meta 87]. In Table 6.16 we show the results of designing a single circuit (Simple OTA) with a 2μ process modelled with BSIM models, a 1.2μ process modelled with BSIM models and a 1.2μ process modelled with MOS3 models. In each of these cases, the performance specifications given are identical and ASTRX/OBLX was told to minimize area. As expected, the 2μ process required the largest design to meet the specifications. Again, the OBLX predictions match the HSPICE predictions almost exactly. The most interesting aspect of this result is that the 1.2μ BSIM and 1.2μ MOS3 designs yield dramatically different areas, yet the model parameters are derived from the same fabrication process. Clearly, the models behave much differently for the device parameters of this circuit, and this highlights the sensitivity of the design to the device models.

As a second example of the utility of encapsulated device evaluators, we include a BiCMOS two-stage amplifier, which shows the ability of ASTRX/OBLX to handle a mix of MOS and bipolar devices. The circuit schematic is in Figure 6.6, while synthesis and simulation results appear in Table 6.17. Note the high-performance of this design and again the excellent agreement between OBLX prediction and simulation.

This BiCMOS design example is important also because of the exponential nature of the current-voltage relationship in a bipolar device. Because of this exponential nature, solving for the D.C. operating point in circuits with bipolar devices is typically more difficult than for MOS only circuits. This fact was a cause for speculation that the relaxed D.C. formulation would not be able to converge to D.C. operating points for circuits that contain bipolar devices. The ability to design this BiCMOS circuit in ASTRX/OBLX is an existence proof that the relaxed D.C. formulation works for both MOS and bipolar devices.

TABLE 6.16 Comparison of Models and Processes for Circuit Simple OTA

	Specification: OBLX / Simulation		
Attribute	**2μ BSIM**	**1.2μ BSIM**	**1.2μ MOS3**
Cload	1pF	1pF	1pF
Vdd	5V	5V	5V
D.C. gain (dB)	≥26: 26.0 / 26.0	26.1 / 26.1	51.2 / 51.2
gain bandwidth (MHz)	≥20: 20.5 / 20.4	20.0 / 20.0	20.0 / 20.0
phase margin (°)	≥60: 88.1 / 88.0	90.5 / 90.3	88.5 / 89.8
PSRR (Vss)	≥20: 23.2 / 23.2	29.4 / 29.4	45.9 / 46.0
PSRR (Vdd)	≥20: 26.6 / 26.6	26.6 / 26.6	51.2 / 51.2
output swing (V)	≥2.2: 3.6 / 3.5	3.3 / 3.8	3.1 / 3.1
slew rate (V/μs)	≥10: 39.7 / 40	52 / 52	51.2 / 50.0
active area ($10^3\mu^2$)	↓: 577	305	140
static power (mW)	≤1: 0.20 / 0.20	0.72 / 0.26	0.26 / 0.26
time/ckt. eval (ms)	26	27	24
CPU time (min. / run)	5	5	5

6.3 Design Space Exploration

Our third set of results focuses on design space exploration, which is the ability to generate a number of circuits that are the same in most respects but differ in one or two specifications. This is one of the most important capabilities of a synthesis tool because it allows the designer to explore various performance trade-offs for the topology at hand, quantifying the interactions between competing circuit performance constraints. With ASTRX/OBLX there are two ways to perform design space exploration. The first method is more straightforward to perform, but less direct. Here we allow the stochastic nature of simulated annealing to explore the design space for us. By setting up the cost function with one or more objective terms and running a large number of designs, OBLX will automatically generate results with different trade-offs among these objective terms. For example, Figure 6.7 is a scatter plot that shows the results of this approach. In the graph, each point represents a *Simple OTA* circuit design generated by OBLX. Since the Y axis is area and the X axis is unity gain frequency, the best designs are towards the lower right-hand corner. The boundary of the region populated with designs (marked as a curve on the graph) shows the

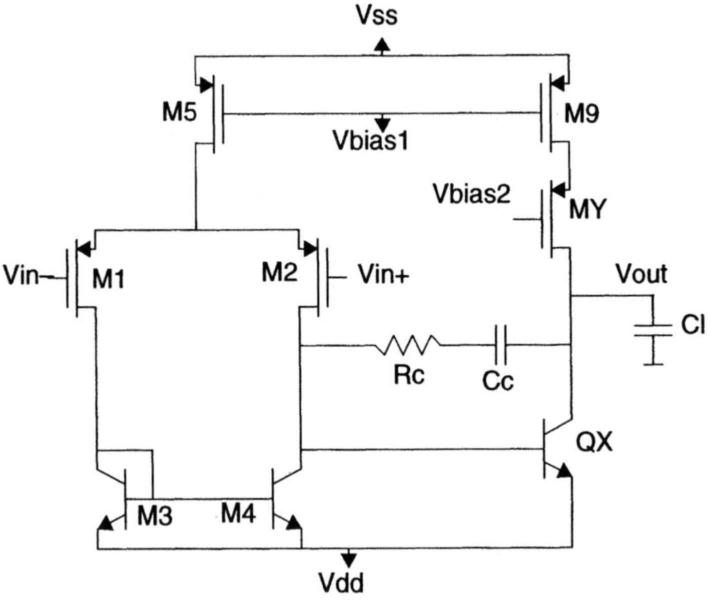

FIGURE 6.6 Schematic Diagram for BiCMOS Two Stage Amplifier

performance limit of this topology given the other specifications. Note that although the circuit designs appear fairly widely scattered, the scale of the graph is quite small and the designs near the top of the graph are also quite acceptable.

The second method of performing design space exploration is more direct. In this approach, the performance parameters of interest are given as specifications rather than objectives. The values for these are varied and new OBLX runs are performed for each set of specifications. The best of these runs is selected as the circuit that represents that group of specifications. This is the approach followed to generate the graph in Figure 6.8, which shows results of several ASTRX/OBLX runs for circuit *Comparator*. Each point on the graph is a different complete circuit design, obtained by increasing the clock frequency specification and asking ASTRX/OBLX to minimize static power. The graph shows the expected increase in static power consumption as a function of the circuit's maximum clock frequency.

TABLE 6.17 BiCMOS Two-Stage Amplifier, BSIM/BJT Models, 1.2μ Process

Attribute	Specification:	OBLX/Simulation
Cload		1pF
Vdd		5V
D.C. gain (dB)	↑:	99.1 / 99.1
gain bandwidth (MHz)	≥50:	73.7 / 75.1
phase margin (°)	≥45:	45.2 / 49.6
PSRR (Vss)	≥60:	78.9 / 79.0
PSRR (Vdd)	≥40:	52.2 / 52.2
output swing (V)	≥2:	3.3 / 4.0
slew rate (V/μs)	≥10:	10 / 9.5
active area ($10^3\mu^2$)	↓:	11.9
static power (mW)	≤20:	1.3 / 1.5
time/ckt eval (ms)		38
CPU time (min. / run)		12

6.4 Comparison with Manual Design

Our fourth set of results focuses on comparing ASTRX/OBLX to manual design. The novel folded cascode fully differential opamp, shown in Figure 6.9, is an important test of the ASTRX/OBLX system. This circuit is a unique case for several reasons. First, it is a large complex circuit not found in any textbook and the performance equations cannot simply be looked up. Second, the performance of the circuit is very difficult to express analytically and as many as 6 poles and zeros may non-trivially affect the frequency response near the unity gain point. Finally, the circuit is a novel high-performance design published in [Nakamura 92] and as such is a significant test for any synthesis tool. Further, we have access to the device sizes originally obtained by an expert designer and the original process parameters for which it was designed. Table 6.18 is a comparison of a re-design of this circuit using ASTRX/OBLX with the highly optimized manual design as originally published for the same 2μ process. Surprisingly, ASTRX/OBLX actually finds a design with higher nominal bandwidth

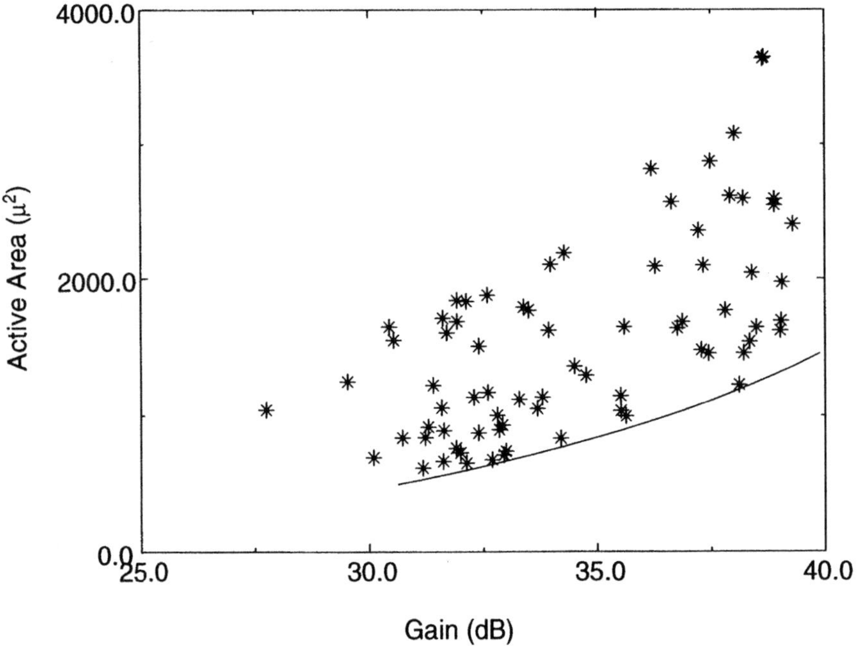

FIGURE 6.7 Exploration of Gain vs. Area for Circuit *Simple OTA*

at the cost of less area. Althougn we are pleased with the ability of OBLX to find this corner of the design space, this does not mean that ASTRX/OBLX out-performed the

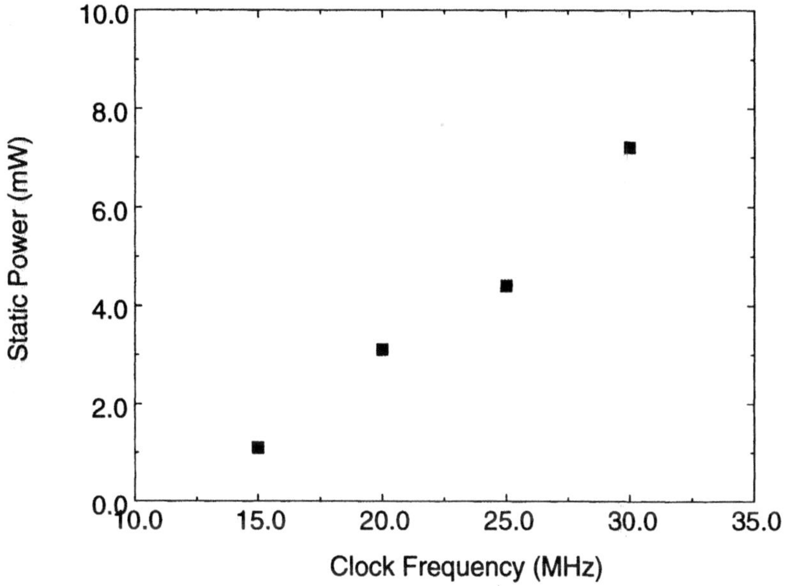

FIGURE 6.8 Exploration of Clock Frequency vs. Static Power Circuit *Comparator*

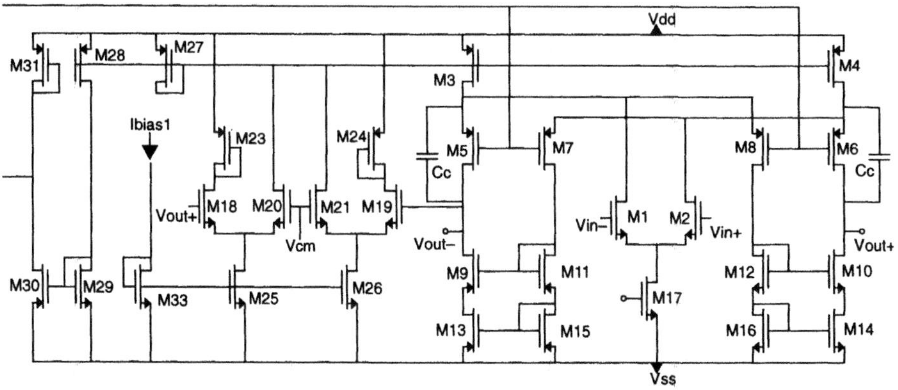

FIGURE 6.9 Novel Folded Cascode Operational Amplifier: Circuit Schematic

manual designer in all respects. In fact, as we shall see when we revisit this example in Section 8.1, the manual designer was willing to trade nominal performance for better overall yield and performance over varying operating conditions. Adding this ability to ASTRX/OBLX is the subject of the second part of this book.

This circuit also highlights a potentially important use of a synthesis tool that has yet to be exploited. Because equation-based approaches require a burdensome amount of preparatory effort, the topologies incorporated into the libraries of equation-based tools must justify the effort it takes to put them there. Thus, it is impractical to use these kinds of tools to evaluate new and experimental topologies. With ASTRX/OBLX such comparisons are straightforward. The new folded cascode circuit in this section is derived from the basic folded cascode circuit that formed part of our comparison suite (see Figure 6.4). By designing both of these circuits with ASTRX/OBLX we can see the effects of the topological modifications. Comparing Table 6.9 and Table 6.18, we see that the changes resulted in improved gain, frequency

TABLE 6.18 Comparison with Manual Design for Circuit Novel Folded Cascode

		Automatic Re-Synthesis with ASTRX/OBLX		
Attribute	**Manual Design**	**Specification**	**Design**	**Simulation**
Cload	1pF	1pF	1pF	1pF
Vdd	5V	5V	5V	5V
D.C. gain (dB)	71.2	≥ 71.2	82	82
gain bandwidth (MHz)	47.8	↑	89.3	89.0
phase margin (°)	77.4	≥ 60	91	91.1
PSRR (Vss)	92.6	≥ 93	112.1	112
PSRR (Vdd)	72.3	≥ 73	77.3	77.2
output swing (V)	± 1.4	± 1.4	± 1.4	1.3
slew rate (V/μs)	76.8	≥ 76	92.5	87
active area ($10^3\mu^2$)	68.7	↓	56.5	
static power (mW)	9.0	≤ 25.0	12.1	12.1
time/ckt. eval (mS)			83	

response and slew rate at the price of increased area and decreased power supply rejection. This kind of direct comparison becomes practical when there is little or no barrier to designing new circuit topologies with a synthesis system.

6.5 Designing Large, Realistic Cells

In this section, with our final set of results, we discuss the utility of the ASTRX/OBLX synthesis methodology when confronted with the complex task of designing a pipelined analog-to-digital converter (ADC). This is an important test because it addresses the issues of *generality* and *complexity*. Previous approaches to analog synthesis focused on small linear circuits and validated their work by synthesizing textbook circuits with a handful of transistors. Here we show that ASTRX/OBLX can automatically synthesize the two sampled-data cells needed to complete the design of the ADC shown in Figure 6.10: a switched-capacitor gain cell and a comparator. These cells display important non-linear behavioral characteristics whose modelling in ASTRX/OBLX is not as straightforward as the performance characteristics of lin-

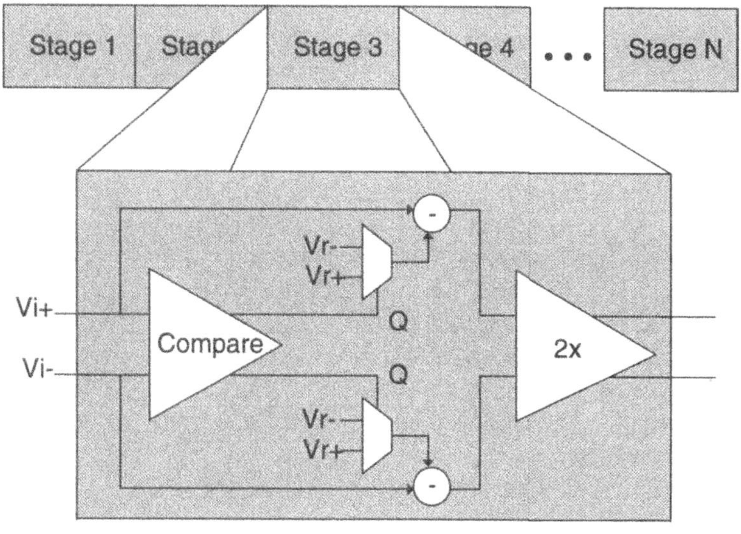

FIGURE 6.10 Pipelined A/D Converter Topology

ear cells. Since the comparator is part of our comparison suite, we focus on completing the pipe with the 2x gain cell. This cell is the largest fully automatic analog cell synthesis result of which we are aware.

The pipelined A/D topology converter we selected (Figure 6.10) employs two cells: the comparator from our comparison suite (Figure 6.5) and a 2x gain amplifier (Figure 6.11). To aid noise rejection in a mixed-signal environment, we use a fully differential structure. The 2x gain cell is implemented as a switched-capacitor circuit, and the input switches provide a convenient method to perform the multiplexing and subtraction needed to complete the design. The 2x gain cell also employs an operational amplifier, and for this we use the novel folded cascode fully differential operational amplifier (Figure 6.9). It is important to note that ASTRX/OBLX optimizes the *entire* gain cell as a *single*, *flat* circuit. This allows the optimizer to explore crucial trade-offs among the sizes of the switches, capacitors and amplifier devices. Unfortunately, this also yields a very large optimization problem, which also contributes to the lack of published synthesis results of this complexity. However, the ability to han-

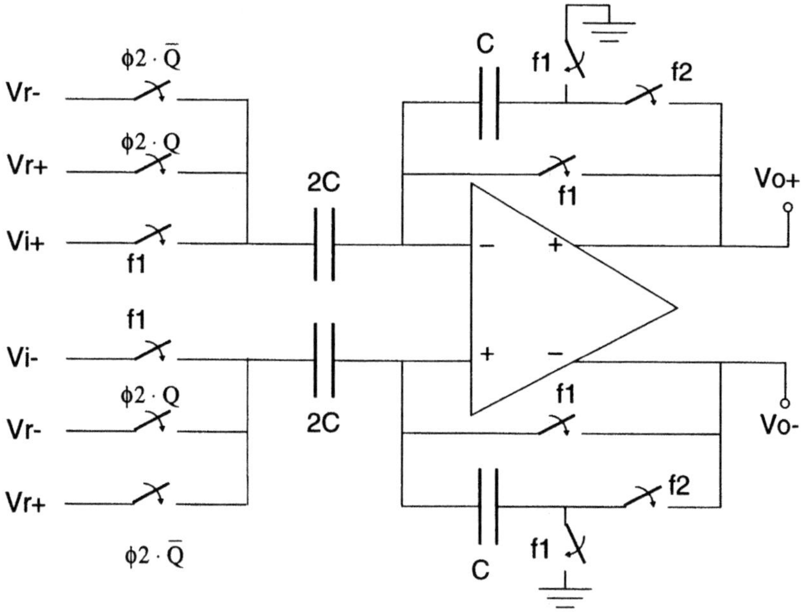

FIGURE 6.11 2x Gain Switched Capacitor Circuit Schematic

dle problems of this magnitude is fundamental for industrial design situations where the designer naturally works with cells of this complexity.

Sample ASTRX/OBLX synthesis results for a 2x gain stage appear in Table 6.12. These results are for a 1.2μ MOS process and use a BSIM model for the devices. Reported run-times are on an IBM RS/6000 550, and simulation results were obtained with HSPICE. Again, note the close correspondence between OBLX prediction and simulation.

Finally, Figure 6.13 shows the simulated input/output response at 3MHz for a single stage of our completed A/D converted. As expected, the curve has a sawtooth shape: for inputs beyond one-half full-scale, one-half full-scale is subtracted before the times two multiplication is made. We consider this perhaps modest circuit performance to be quite acceptable for the first fully automatic design of cells of this complexity.

6.5.1 The Effect of Additional Variables/Constraints on Run-Time

As the size of the cells we synthesize increases, it is important to be able to predict how much more computational power will be required to design the cell. Accurate prediction is quite complex and depends on many details of the problem. There are two major determining factors in the overall run-time of the optimization: the amount

FIGURE 6.12 Swit-cap 2x Gain Amp Synthesis Results

Key Attribute	Specification:	OBLX/Simulation
Vdd	5V	
settling at output (ns)	≤100:	93 / 98
settling at amp inputs (ns)	≤100:	42 / 61
input range (V)	≥±0.5:	1.6 / 1.6
output range (V)	≥±0.5:	0.8 / 0.7
common mode gain (dB)	≤-10:	-33 / -33
active area ($10^3\mu^2$)	minimize:	58
static power (mW)	minimize:	21 / 21
time/ckt eval (ms)	158	
CPU time (hrs. / run)	11.8	

of time to evaluate a single move and the number of moves. These factors are in turn influenced by other problem details. For example, evaluating a single move may require only one AWE analysis, or may require several. For each AWE analysis, the number of calls to non-linear device models, the number of nodes in the circuit, and the requested order of the approximation all influence the evaluation time. The number of moves required to find a good solution will in turn be influenced by the number of independent variables and the intrinsic "difficulty" of the problem, *i.e.*, how difficult it is to search for parameter settings that meet a set of aggressive, mutually incompatible design goals. In this section we attempt to explore and quantify some of these influences on overall design time with ASTRX/OBLX.

We begin by completing the ASTRX analysis information discussed for the comparison suite in Section 6.1. Table 6.19 lists this information for the other three circuits discussed in this chapter. Together, these eight circuits will give a fairly good overall look at how run-time scales. The information of most relevance to run-time has been collected in Table 6.20. Run-time is simply the product of the number of moves and the average time per move. Recall that by default OBLX uses the heuristic formula in

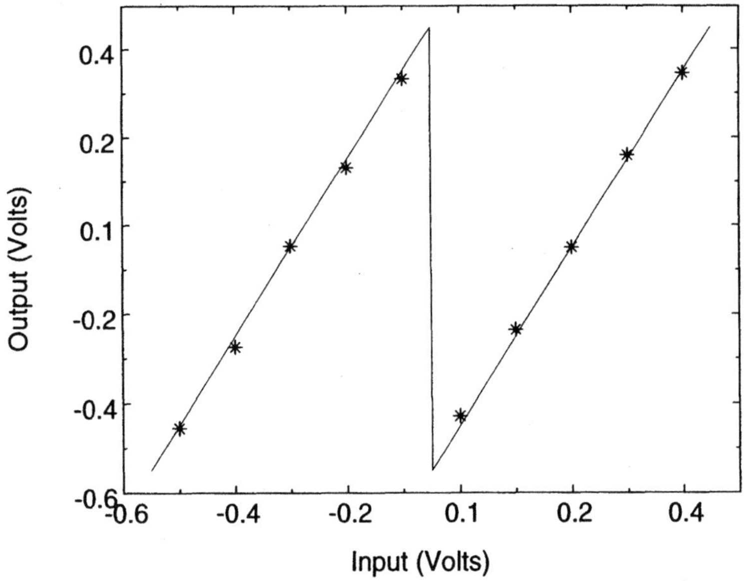

FIGURE 6.13 Input /Output Response of Pipeline Stage

TABLE 6.19 Result of ASTRX's Analyses for Other Circuits

		Circuit		
		BiCMOS 2-stage	Novel Folded Cascode	2x Gain Stage
Input (lines)	Netlist/Models	39	68	189
	Synth. Specific	33	51	65
$\underline{x}$	User-Supplied	12	27	39
	Node Voltages	26	84	215
$C(\underline{x})$	Terms	86	246	483
	Lines of C	1723	3960	7173
Circuits	Circuit Type[a]: nodes, elements	B: 33, 54 A: 32, 105	B: 90, 167 A: 90, 395	B: 219, 450 A: 152, 693 A: 152, 693 A: 90, 395

a. Type 'A' is a linearized AWE circuit. Type 'B' is a bias circuit.

TABLE 6.20 Collected Run-time Information

Circuit	AWE ckts.	Non-linear Devices	Transfer Functions	Nodes in Bias Ckt.	Total AWE Nodes	Time/Ckt. Eval (ms)	Variables	Est. Moves/ Run (X1000)	Run-time (min.)
Simple OTA	1	5	3	20	20	41	21	7	7
OTA	1	9	3	28	29	37	35	14.5	10
BiCMOS 2-stage	1	8	3	33	32	35	38	16	12
Two-Stage	1	9	3	34	33	38	45	21	14
Novel Folded Cascode	1	32	3	90	90	83	111	75	116
Folded Cascode	2	26	5	75	150	116	98	61	120
Comparator	3	23	4	64	156	136	76	43	97
2x Gain Stage	3	83	4	219	394	158	234	207	708

Equation 6.1 to estimate the number of moves that will be required to find a minimum.

$$\text{Moves} = K\|\underline{x}\|^{\eta} \tag{6.1}$$

where K defaults to 100 and η to 1.4. Thus, by default, the number of moves is superlinearly related to the number of independent variables. For very difficult problems, this number could be increased, although it is typically not necessary.
Figure 6.14 shows the effect of increasing the number of variables on run-time. Note that although the graph seems quite consistent for circuits with a single AWE evaluation (the run-time increases as about $n^{1.7}$ for n variables), the run-times for the more complex circuits seem inconsistent. Clearly, run-time increases significantly for circuits that require more than one AWE evaluation because more time is spent per move, but this increased time is difficult to estimate. For example, for both the comparator and the 2x gain cell, the circuits evaluated by AWE vary in size. Where possible, smaller sub-circuits are used to evaluate some of the performance specifications. Thus, the amount of time spent in AWE is not simply three times as much because of the three evaluations. Further, the amount of time required for each move varies with

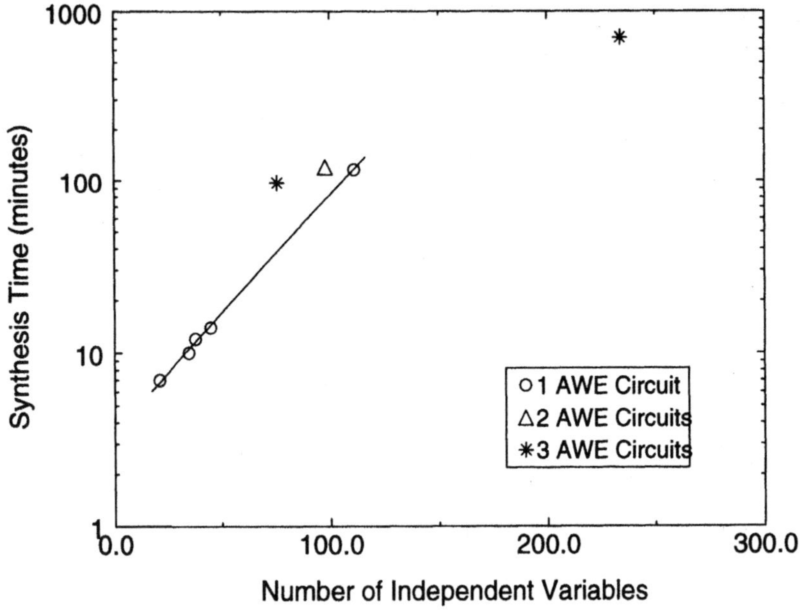

FIGURE 6.14 Number of Independent Variables vs. Run-time

other important influences such as the number of transfer functions needed and the number of non-linear device model evaluations required. For small circuits, the sparse-matrix LU-factorization step within AWE is dominated by the dense matrix manipulation required to complete the Padé approximation. Further, generating the linearized elements can be quite time-consuming. For example, a single call to a BSIM model still takes nearly a millisecond on a 60 MIP workstation. Overall, predicting run-time is not a straightforward task because of the many complicating factors. However, we can say that for a given number of required AWE analyses it does increase superlinearly—but not quadratically—with the number of design variables.

We have also performed a second study, this time of the relationship between run-time and the number of nodes in the circuit under design. Because each of our design examples requires only a single bias circuit, the number of nodes in the bias circuit was selected as an indicator of problem complexity. The graph in Figure 6.15 depicts the relationship between run-time and number of nodes in the bias circuit for the eight circuit synthesis results in this chapter. As with the relationship between the number of variables and run-time, the trend is quite clear for circuits with a single AWE cir-

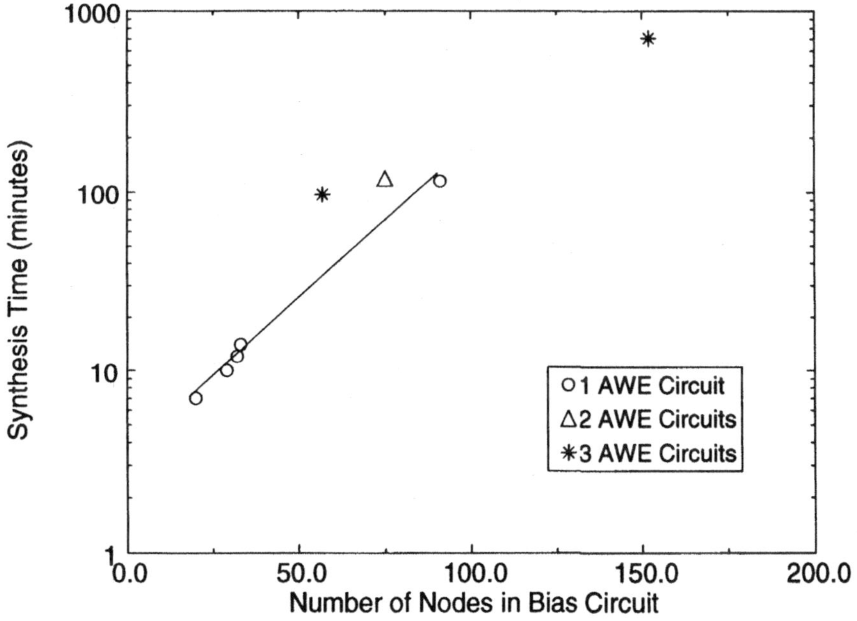

FIGURE 6.15 Number of Nodes in Bias Circuit vs. Run-time

cuit: run-time increases as about $n^{1.8}$, where n is the number of nodes in the bias circuit. As described for the relationship between variables and run-time, if more AWE analyses are required to evaluate each circuit configuration, the relationship is more complex. Nevertheless, as before, given a number of required AWE analyses, run-time increase is superlinear, but somewhat less that quadratic, with the number of nodes in the bias circuit.

6.6 Summary

In Section 1.4 we defined our goals for a nominal synthesis tool in terms of the metrics accuracy, automation, and breadth. In Chapter 2, we compared previous approaches using these metrics. In this chapter we presented a comprehensive suite of circuit synthesis results for ASTRX/OBLX and described how these results showed significant progress towards our goals and clear improvement over prior approaches when evaluated with the same metrics. In the next chapter, we present further synthesis results but examine them from a different perspective, verifying the basic soundness of the engineering decisions that went into the design and implementation of the ASTRX/OBLX framework.

CHAPTER 7

Validating the Tool Design

In the previous chapter we focused on nominal circuit synthesis results from the perspective of an analog designer, concentrating on circuit examples and comparing the capabilities of the tool with those of previous tools. In this chapter we focus on synthesis results from a CAD tool designer's perspective, concentrating on the decisions made in the design and implementation of the system and showing how effective they are. We begin with a discussion of the design strategy used in constructing the annealing control mechanisms such that the number of "tuning constants" that need to be adjusted by a user is minimal. We then examine the more important mechanisms built into the OBLX annealing library: the cooling schedule, the dynamic move class selection biasing, and the dynamic weighting. Next, we revisit the relaxed D.C. assumption in our optimization formulation and look at how D.C. correctness evolves during the annealing process. Finally, we discuss the overall annealing process and the evolution of a circuit design as the various terms in the function are minimized.

7.1 Reducing the Need for User-Controlled Constants

An important concern in the usability and automation of ASTRX/OBLX is reducing the number of constants that must be adjusted by a user in order to complete a circuit design. In optimization there are often weights, scale factors, and other "magic numbers" that are required for proper numerical performance of the algorithm. A tool does not actually automate design if it transforms a complex circuit problem into a complex optimization problem, and then requires the user to select myriad constants that require specialized knowledge of the tool's implementation. In this case the circuit design problem has *not* been solved—merely converted into an equally difficult optimization problem. However, in any complex numerical algorithm some constants are required. Our goal then is to minimize the amount of "constant adjustment" that must be performed by a designer when using ASTRX/OBLX to design a circuit. We address this goal with two strategies:

1. We design the internal components of the annealer to be robust across multiple circuits synthesis problems.
2. We "tune" the remaining constants such that we obtain the best possible performance across various different synthesis problems.

These two steps are difficult to accomplish and to verify. Because of the difficulty of the problems, complexity of ASTRX/OBLX, and time required to complete a design, finding the "optimal" set of mechanisms and tuning them to the "optimal" constant values for all possible designs is simply unrealistic. Hence, we set a more modest goal: an analog designer new to ASTRX/OBLX should expect a reasonable solution to a correctly specified synthesis problem without adjusting any constants. Simulated annealing is a stochastic process, and each time OBLX is run it will find a slightly different trade-off among its constraints. As a result, it is not possible to guarantee that a single run will provide the best answer. However, we wish to be confident that running OBLX several (5-10) times will provide several high-quality solutions from which to choose. We discussed the first of the two strategies aimed at achieving this in Chapter 4. The cost function and annealing control mechanisms are designed to be as robust as possible. In particular, an automatic cooling schedule, dynamic move class selection biasing, and dynamic weighting scheme were incorporated into our annealing formulation to improve this robustness. We present results to quantify the performance of these mechanisms in Section 7.2, Section 7.3, and Section 7.4, respectively. In this section, we discuss the second strategy: the method we used to tune the constants and the results of this tuning process.

For any numerical algorithm, there is typically a tuning process when the constants in the algorithm are adjusted to achieve the best overall performance. Normally this process is performed by the algorithm's designer or by someone else intimately familiar with its details. Normally this manual tuning is performed using the following informal steps:

1. Based on intuition and expert knowledge of the algorithm, attempt to select constants that work.

2. Run several examples through the numerical algorithm and note its performance. If the performance is acceptable, quit.
3. Based on intuition, expert knowledge, and the observations of the gathered from running examples, attempt to adjust the constants so that the algorithm performs better.
4. Go to Step 2.

These steps work well when the number of constants is small and running several examples is quite quick. However, when the number of constants is substantial and examples run more slowly, as is the case with ASTRX/OBLX, a more sophisticated technique is called for. Originally, we used the simple steps above to select our constants, relying on our insight into the algorithms to guide us to constant values that would be effective. However, as the annealing mechanisms we employed became more complex, the number of constants and their interactions became overwhelming. What became apparent during this manual tuning process was that the process of selecting constants to control the annealing mechanisms was itself an optimization process, where the variables in this optimization are the constants used by the annealer. With this realization, it then became apparent that the best method of selecting the constants to optimize the performance of OBLX was to use a formal optimization technique. We refer to this process as *optimizing the optimizer*.

To define completely how we optimized the optimizer, we need to describe it as an optimization algorithm. As with the description of the annealer that forms the core of our nominal synthesis formulation, we describe the variables the optimizer manipulates, the cost function we are minimizing, and the algorithm used for minimization. Clearly, the variables used for optimizing the optimizer, are the parameters for the annealing control mechanisms. Because calling constants variables is somewhat confusing, we shall denote these as $\underline{\gamma}$. Note that not all the constants used by OBLX necessarily end up in $\underline{\gamma}$, constants that refer to physical properties—such as tolerances on voltages and currents—should be set once and changed only for problems with different physical tolerances. In general, the constants in $\underline{\gamma}$ are those that affect the annealer's dynamic weight scheme, move-set, and cooling schedule. In Table 7.1 , we categorize the OBLX parameters in $\underline{\gamma}$. In total, there are 56 parameters in $\underline{\gamma}$, which makes this a fairly large optimization problem.

The next step in describing the optimizing-the-optimizer process is the cost function. In designing a suitable cost function, we must determine a quantifiable performance metric, that we can formulate as a cost function, $\Psi(\underline{\gamma})$, that measures the performance of ASTRX/OBLX. As the first step in defining this cost function, we define an auxiliary function, $\chi(\underline{\gamma})$, that evaluates a single annealing run, as:

$$\chi(\underline{\gamma}) = \chi^{obj} + \chi^{con} + \chi^{dev} + \chi^{dc} + \chi^{t} \quad (7.1)$$

where we define the terms as:

- χ^{obj} is the final value of C^{obj}, *i.e.*, the final value of the objective terms from the annealing run.
- χ^{con} is the final value of the errors in the performance constraints, weighted by a scaling factor. This is a measure of the how well the performance constraints were met by this run. Note that this is related to, but not exactly equal to the final value of C^{con} (see Section 4.4.2). Since C^{con} includes weight terms that were set dynamically during the anneal, using it might bias our cost function so that it min-

TABLE 7.1 OBLX Parameters in $\underline{\gamma}$

Parameter Category	Number of Variables in $\underline{\gamma}$
Trajectory for C^{KCL}	8
Other for C^{KCL}	5
Trajectory for C^{dv}	8
Other for C^{dv}	3
Trajectory for C^{CON}	8
Other for C^{CON}	3
Trajectory for C^{DEV}	8
Other for C^{DEV}	3
Percent Moves Under Hustin Control	1
Newton-Raphson Move Class Generation	7
Time to Start Contracting Move Ranges	1
Lam Total Move Attempts Control	1

imizes the weights used during annealing, rather than minimizing the errors. Formally, using the notation of Section 4.4.2, we can write χ^{con} in terms of the final value of the annealing variables as:

$$\chi^{con}(\underline{x}) = \frac{1}{\left\|\Omega^{con}\right\|} \cdot \sum_{i \in \Omega^{con}} \max(\hat{g}_i(\underline{x}), 0) \tag{7.2}$$

This term should be zero for a successful run, *i.e.*, no performance constraints are violated by the final solution. We scale χ^{con} by a weight, 5 in our case, that is designed to increase the importance of driving the term to zero.

- χ^{dev} is similar to χ^{con}, except that it is formulated from the device region of operation terms in the cost function and is weighted by the constant 100. The higher weight reflects the smaller typical error values in device terms.
- χ^{dc} is also similar to χ^{con}, except that it is formulated from the D.C. operating point terms and is weighted by the constant 1000. In this case, the higher weight reflects the notion that it is most important to drive this term to zero.
- χ^{t} is not derived from the final circuit solution determined by the annealer. Instead, χ^{t} is simply the number of CPU seconds required for the anneal multiplied by a scaling weight. This term reflects the fact that we wish the annealer to be efficient and to produce high-quality circuit solutions. The scaling weight depends on the typical run-times for the circuit problem we are synthesizing with ASTRX/OBLX. We set this weight so that χ^{t} is a small fraction of χ^{obj} because quality is more important than efficiency for our application. In the experiments discussed below, this weight was 0.001.

Because of the stochastic nature of annealing, it is insufficient to evaluate $\Psi(\underline{\gamma})$ as a single $\chi(\underline{\gamma})$, *i.e.*, the performance of a set of parameters should not be based on a single OBLX run. Instead, we must run OBLX multiple times and use the statistics from the complete set of runs. Ideally, each time we modified the annealing constants, we would run OBLX *many hundreds* of times on *every* circuit problem we have, ensuring that we had an accurate measure of the effect of the modification. Unfortunately, this is computationally infeasible. Given that we are optimizing 56 variables, we expect to perform at least several hundred moves, depending on the optimization method we choose. A single OBLX run with our smallest circuit takes 5 minutes of CPU time, but a single run for *all* the circuit problems in chapter 6 takes over 18 hours. Assuming we can run all the annealers required for a statistical sample in parallel across a network of machines, optimizing the optimizer over all these circuit problems would take at least several years to complete. Because we have more modest resources and are not willing to wait that long, we decided to optimize only over a single circuit, our smallest. Other than practicality, our rationale for this choice was the belief that we had designed the annealing cost function and annealing control mechanisms to be robust across different synthesis problems. Given the decision to design only our smallest circuit, circuit *Simple OTA* (see Section 6.1.1), we then needed to determine the number of runs necessary for a statistical sample. Here, again, long run-times were the strongest influence in the decision. A few simple experiments showed that there was substantial variance in the statistical properties of $\chi(\underline{\gamma})$ applied to fewer than 200 OBLX runs. As a result, we select 200 as our sample size. Having fixed the sample size, we then formulate $\Psi(\underline{\gamma})$ as the sum of the mean, median, and minimum values of $\chi(\underline{\gamma})$ in our sample. Although this is somewhat redundant, the combination of mean and median is intended to keep the distribution somewhat narrow. Because a few values of $\chi(\underline{\gamma})$ tend to be very large, and we were willing to allow a few bad designs for better overall quality, the variance was found to be an inappropriate measure.

To speed the process of performing 200 annealing runs, we have implemented a system that distributes the execution of these multiple OBLX runs over a computer network. This tool can employ the idle time of a network of over 200 workstations throughout Carnegie Mellon. In practice, we enjoy speedups of 20-50 over serial execution.

The final aspect of the optimizing-the-optimizer formulation is the optimization algorithm itself. We employ a simple variation of Powell's method of optimization [Powell 72, Powell 75] due to [Press 88]. Since this is a fairly standard technique, we direct the interested reader to [Brent 73] for a more detailed treatment. For our application, the salient features of the algorithm are that it is a local optimization technique which does not require derivatives. Given that OBLX has simulated annealing as its core optimization algorithm, the selection of a local optimization technique may seem somewhat surprising. However, as with global optimization techniques in general, simulated annealing typically requires many more function evaluations than a local optimization technique for a comparable problem. Since function evaluations take several hours, this reduction in function evaluations is crucial. Another important advantage of local optimization techniques over simulated annealing for optimizing the optimizer is that the local techniques improve the value of the cost function at each step. Because a typical optimizing-the-optimizer run requires several months, we may wish to (or have to) terminate the run prematurely. With a local technique, we are assured that any intermediate solution is at least as good as our initial starting point. Finally, because the initial values for $\underline{\gamma}$ were chosen after careful study of the algorithms, we believe that the final value for $\underline{\gamma}$ should lie near enough to this initial point that a global minimization technique is not warranted. The second important characteristic of Powell's method is that it does not require derivatives. Clearly, analytic derivatives of $\Psi(\underline{\gamma})$ are not available. Computing derivatives numerically is costly in terms of function evaluations. Further, because $\Psi(\underline{\gamma})$ is statistical in nature, it is a very "rough" function and numerical differentiation might very well yield wildly inaccurate results. Because Powell's method is a well understood local optimization algorithm that does not require derivatives, it was selected as the core algorithm for optimizing the optimizer.

After several false starts, we successfully ran our optimizing-the-optimizer algorithm for an extended period of time. Because of time constraints, we did not allow the optimizer to complete but stopped it when it seemed quite close to converging. This took several months of real time, and approximately 4 years of CPU time. The most important results of this process are the set of trajectories for the dynamic weight mechanism shown in Figure 7.1. (see Section 4.5.2 for a description of the method in which these trajectories are used.) These trajectories are interesting because they are somewhat different than the trajectories *we* originally designed. In particular, the trajectory for the C^{KCL} weights has a sharp peak near the end of the cooling schedule. When designing these trajectories originally, our trajectory was decreasing rather than increasing at this point. Our design reflects the intuitive notion that we should be forcing the annealer to converge to a D.C. correct solution at this late stage in the annealing process. After studying the trajectories determined by optimizing the optimizer, we believe this peak does make sense. It occurs soon after the cooling schedule has started to force the solution into the bottom of a valley in the cost surface. Because of the tightening of the other constraints, we believe the extra "room" in the C^{KCL} trajectory is necessary to allow moves that improve the other terms but temporarily increase the C^{KCL} term. We regard the fact that optimizing the optimizer found an unanticipated trajectory as an indication of the importance of the method. Note that this kind of unanticipated result is analogous to results reported in [Strenski 91], where annealing parameters were themselves annealed in an attempt to find an optimal cooling schedule, and several surprising results were obtained.

Further results of optimizing the optimizer are shown in Figure 7.2. The histogram on the left shows the starting point for our tuning run. This histogram plots $\chi(\underline{\gamma})$ for 200 synthesized circuits generated by running OBLX on the same design and the same parameters but with 200 different seeds. The height of the left-most bar reflects the number of excellent designs generated. (Note that the complete histogram decays exponentially, but the right-most bar includes all the designs that would not otherwise be shown because of the x-axis scale.) Here the number of excellent designs has improved from roughly 30% to 95% of the total designs. Recall that this took several months of real time. During this time ASTRX/OBLX evolved somewhat, and some of the constants that were determined during the optimization process no longer exist in the most recent version. It was unclear whether these modifications would negate this four month run. Although repeating the tuning run was conceptually straightforward, we would have been required to wait for it to complete. To avoid this additional wait, we performed a simple experiment. We mapped the parameters in $\underline{\gamma}$ to the parameters in the most recent version of OBLX. We then generated histograms by running 200 OBLX jobs, both with the initial values in $\underline{\gamma}$ before optimizing the optimizer and with the final values in $\underline{\gamma}$ after optimizing the optimizer. The result is shown in Figure 7.3. Note that the distribution shows the same progress from the original constants (on the left) to the ones determined by optimizing the optimizer (on the right). From this experiment, we concluded that although the annealer had evolved since the optimizing-the-optimizer run began, we did not need to repeat the run. The remainder

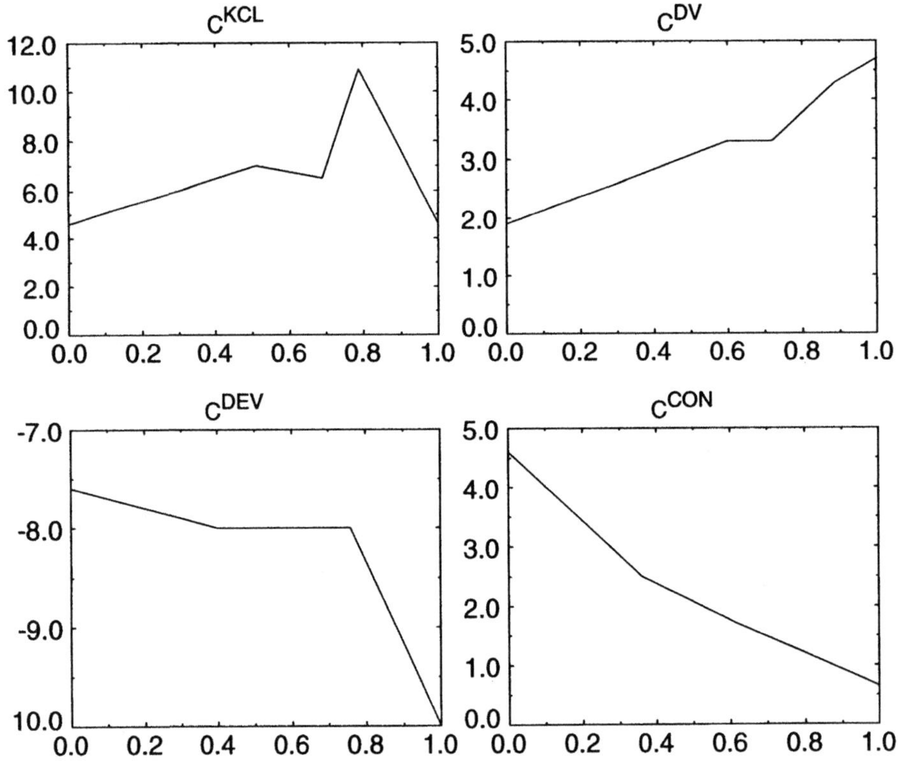

FIGURE 7.1 Actual Trajectories Used For Dynamic Weight Control

of results in this section are generated using the same annealing parameters as those used to generate Figure 7.3.

After showing that optimizing the optimizer improved the value of $\Psi(\underline{\gamma})$ for a single circuit, the next step was determining whether we had obtained improved performance for other circuit problems. Because the cost functions generated by ASTRX are carefully normalized to have the same general characteristics regardless of the circuit and specifications, we expected that the improvement trend seen for this small example would hold for other circuits and specifications as well. To confirm this, we generated several additional histograms for different design problems. For each design problem, we present a pair of histograms. The one on the left was generated

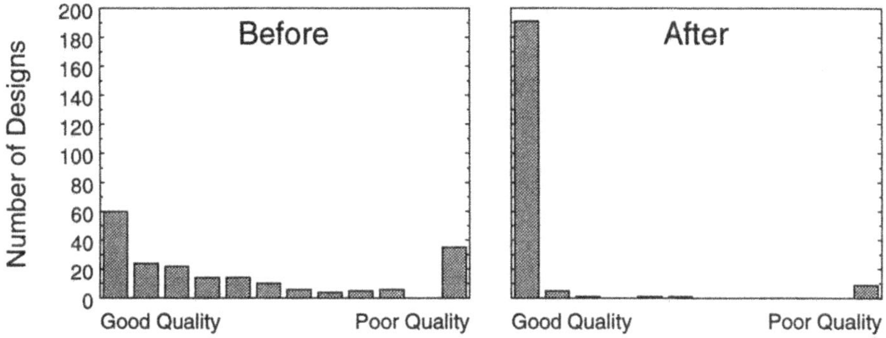

FIGURE 7.2 Before and After Histograms for Optimization of Annealing Constants. (Early Version of ASTRX/OBLX, Circuit *Simple OTA*)

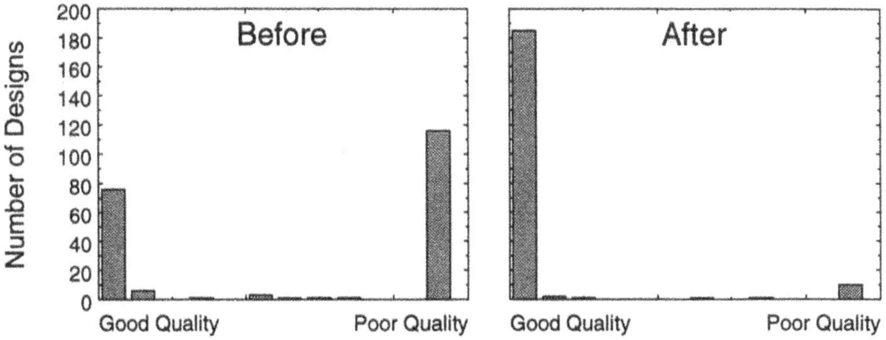

FIGURE 7.3 Before and After Histograms for Circuit Simple OTA (Latest Version of Code)

with the original constants before optimizing them, while the one on the right was generated with the constants at the end of the optimization process. Both histograms were generated with the latest version of ASTRX/OBLX. The first design problem is shown in Figure 7.4. This is the same circuit used for optimizing the optimizer, but the design problem has been made more difficult by tightening the specifications. Note that the histogram on the right is roughly unchanged from that generated for easier specifications (Figure 7.3), but the *before* histogram is significantly worse.

To complete our evaluation of the optimization of the constants used by OBLX, we present histogram pairs for three other circuits. Figure 7.5 is for circuit *OTA*, Figure 7.6 is for circuit *Comparator*, and Figure 7.7 is for circuit *BiCMOS Two-*

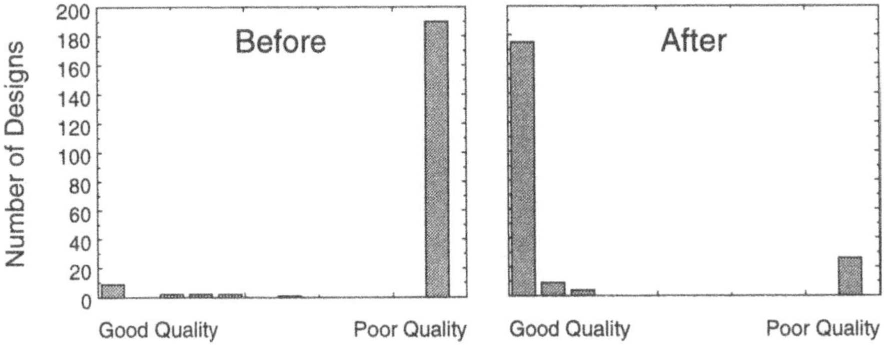

FIGURE 7.4 Before and After Histograms for Circuit *Simple OTA*. (Latest Version of ASTRX/OBLX, More Difficult Specifications)

FIGURE 7.5 Before and After Histograms for Circuit *OTA*

Stage. For all of these circuits, there is improvement in the histograms for the set of circuits generated with the new constants. Note that this improvement trend holds even for the Comparator, which is significantly more complex and modelled with much different techniques than the amplifiers. From this it is clear that the design of the annealer is acceptably robust over different synthesis problems and the process of optimizing OBLX's constants is effective.

However, it is also clear that the dramatic improvement seen in most of the histograms is not shared for the circuit BiCMOS. By examining these circuit results in more detail, we can suggest some explanation for the discrepancy. For the BiCMOS two-stage, D.C. bias points are difficult to determine. The Newton-Raphson moves

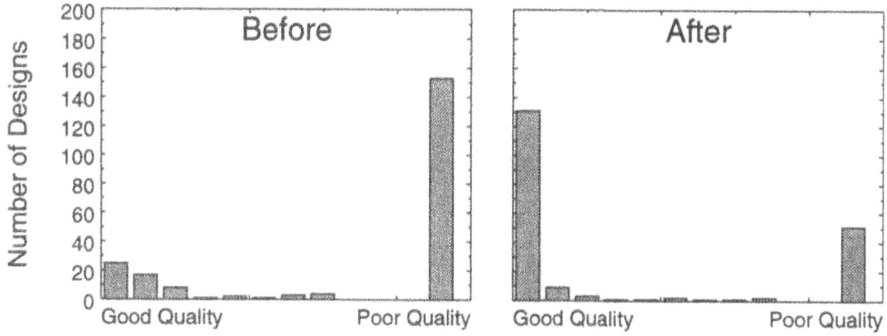

FIGURE 7.6 Before and After Histograms for *Comparator*

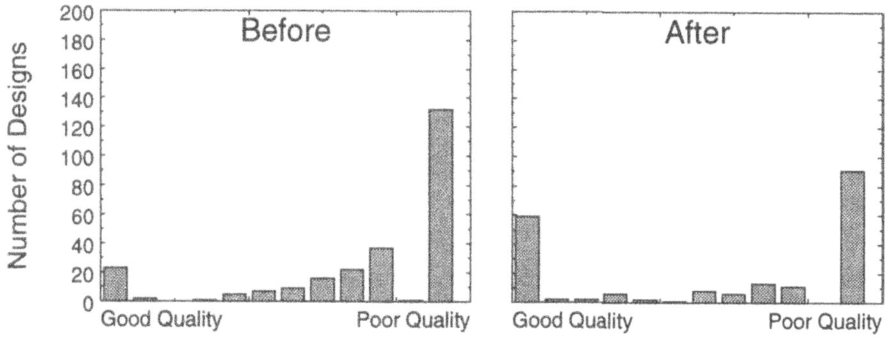

FIGURE 7.7 Before and After Histograms for *BiCMOS Two-Stage*

require significantly more steps to find the operating point than for the other circuits and often fail altogether. Because of this difficulty, substantially more effort should be spent on keeping close to D.C. correctness for this circuit than for the OTA or Comparator. This suggests that a different set of constants may be needed for circuits with special problems of this sort. Of the circuits we have designed, only two circuits (Folded Cascode and this BiCMOS Two-Stage) have this difficulty. Because SPICE3 displays the same D.C. convergence problems as OBLX—while HSPICE does not—we think it likely that the problems stem from device modelling difficulties that have been eradicated in HSPICE. Regardless, we do not see this as a problem that is a priority within the context of ASTRX/OBLX research.

In this section we have developed an effective optimization-based strategy for tuning the numerical parameters that control our annealer. We have shown that this strategy is effective because it has improved the likelihood that a given OBLX run will generate a high-quality circuit solution and is robust across different circuit synthesis problems. In the next three sections, we examine three specific parts of the annealing formulation designed to be controlled by automatic mechanisms: the Lam cooling schedule, the dynamic move-set selection biasing, and the dynamic weights.

7.2 The Lam Cooling Schedule

In this section we examine the effectiveness of our annealer's cooling schedule. Recall from Section 4.5.1 that we use a cooling schedule derived from the efficient λ-schedule. For OBLX, the temperature is controlled by a simple feedback control mechanism. The mechanism monitors the accept ratio, which is the ratio of the number of accepted moves to the number of generated moves. The feedback control is used to adjust the temperature such that the accept ratio follows the target trajectory shown in Figure 7.8. The actual accept ratios shown in the figure are taken from a typical run with the circuit *Simple OTA*. Because an annealing system can be very dynamic, the actual accept ratio curve shows substantial fluctuations. This is to be expected because it is not very effective to treat a problem this small statistically. For very large problems, the statistical behavior is easier to predict and the fluctuations will be less apparent. Regardless, note that these fluctuations are much less apparent on the plot of temperature for the same run, shown in Figure 7.9. The temperature plot shows the typical behavior of any annealing schedule: a slow, controlled cooling process. Note that the Lam schedule has the desirable characteristic that it maintains the temperature at a "simmering" point throughout much of the annealing process.

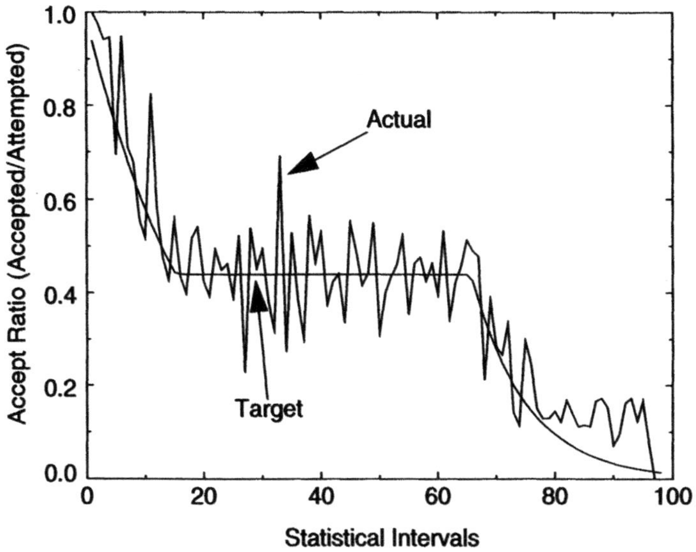

FIGURE 7.8 Move Accept Ratio During Annealing: Lam Target and Actual

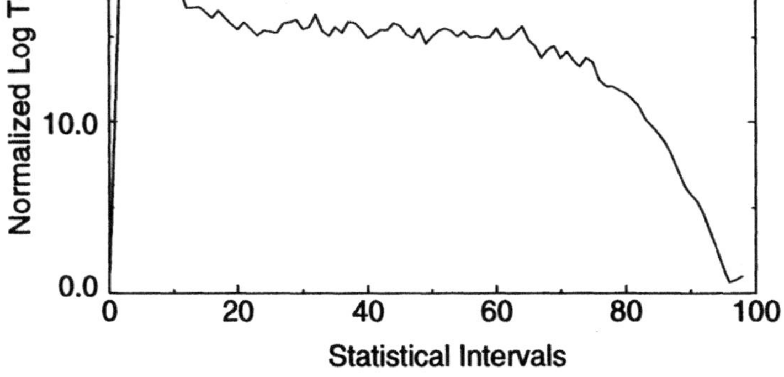

FIGURE 7.9 Temperature During Annealing

7.3 Dynamic Move Selection

The second automatic mechanism built into the annealer is for dynamic control of the move-set. Recall from Section 4.3 that the possible perturbations of the circuit design problem are divided into move classes. The annealer selects from among these classes each time it attempts a move. To improve efficiency, we employ a mechanism that adjusts the probability that a move class will be selected based upon its recent performance. The effect of this mechanism can be seen in Figure 7.10, which was generated from the same annealing run as the curves in Figure 7.8 and Figure 7.9. In Figure 7.10, for a single variable, four move classes were used by OBLX. The figure shows how the probabilities of selecting these move classes vary over the course of the annealing run. The X-axes plot the number of times the move probabilities are updated, which corresponds to the passage of time over the annealing run. The frequency of updates is an annealing parameter that was selected by optimizing the optimizer. The Y-axes are the probability that a given move class will be selected. Note that these probabilities do not sum to 100% because data for other move classes is not being shown. However, the algorithm adjusts between *all* the move classes, trading off amongst move sizes and variables being affected.

The top graph in Figure 7.10 is a composite showing all four move classes. For additional clarity, the four traces are also shown separately in the smaller graphs. From these, we can see that the automatic mechanism behaves much as we theorized. Early in the annealing process, large moves are favored so that a large region of the cost surface can be explored. As the problem cools, these large moves are accepted less frequently and the algorithm increases the probability of selecting medium sized moves. Finally, when the problem is quite cool, only small moves are being accepted and these get the highest probability. Of these graphs, the most interesting are for the small moves. The graph on the lower right is a compound small move, which means that the move is followed by a Newton-Raphson D.C. solve. This explains the two early peaks in this last graph. At these peaks, the move class's probability is increased not because of the small move itself, but because it moves all the node voltages closer to D.C. correctness. Late in the annealing process, the probability reaches its maximum level (10%). Here the KCL terms are being forced to near zero, so only moves that perform Newton-Raphson after adjusting another variable can be very effective. In contrast, the small moves without Newton-Raphson continue to have a low probability late in the annealing process. These graphs show almost ideal behavior for the move selection biasing mechanism.

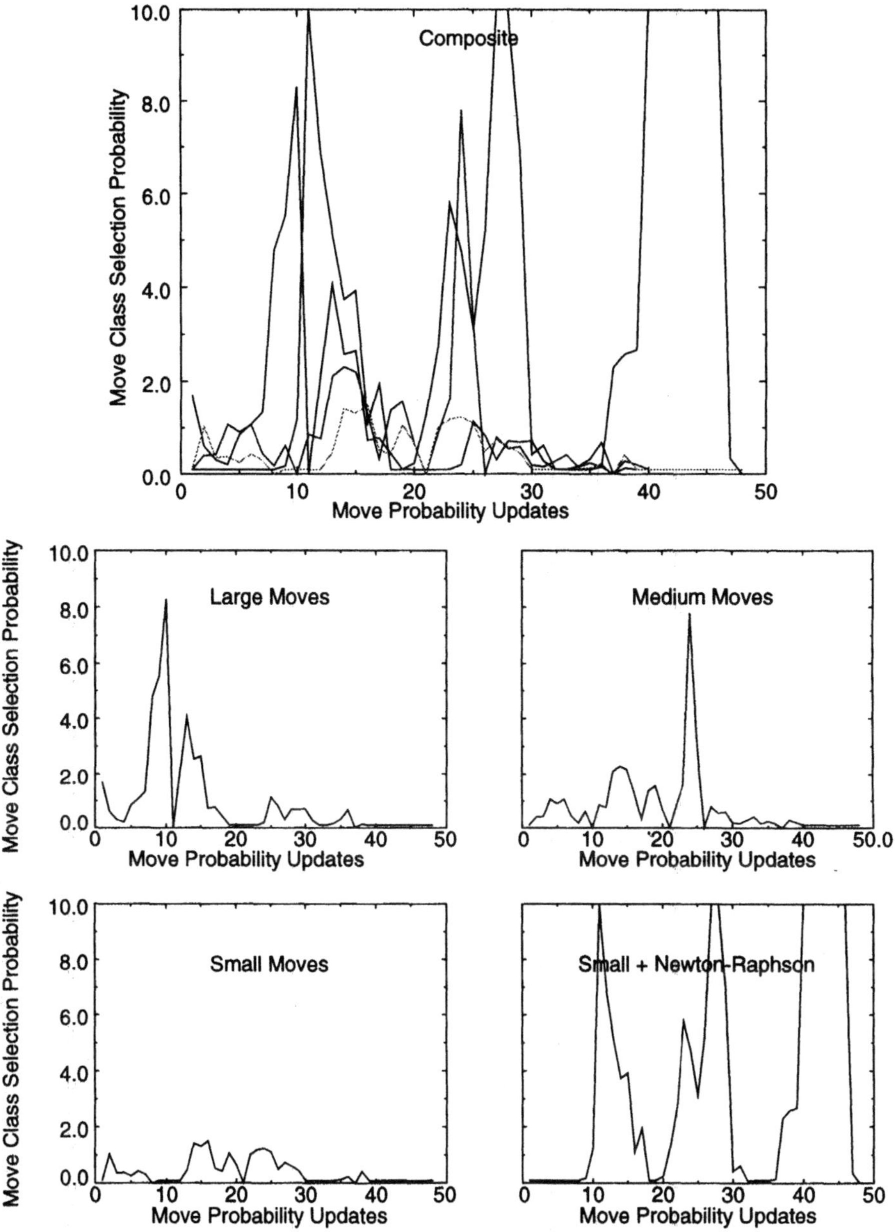

FIGURE 7.10 Dynamics of Automatic Move Class Biasing

7.4 Dynamic Weighting

The final mechanism used to improve annealing robustness is dynamic weighting. With dynamic weighting, we free the user from having to select weights for each of the terms in the cost function generated by ASTRX. This is a significant improvement in automation as there may be several hundred terms for a complex circuit. Recall from Section 4.5.2 that the dynamic weight mechanism uses error trajectories to control the weight. The mechanism compares measured error with a target error and adjusts the weight accordingly. Figure 7.11 shows this process for a collection of different terms. (Note that using the terminology from Section 4.4, $C^{DC} = C^{KCL} + C^{DV}$. Also recall that the shape of the target trajectories shown in these graphs was determined by optimizing the optimizer, as discussed in Section 7.1.) Essentially, if the

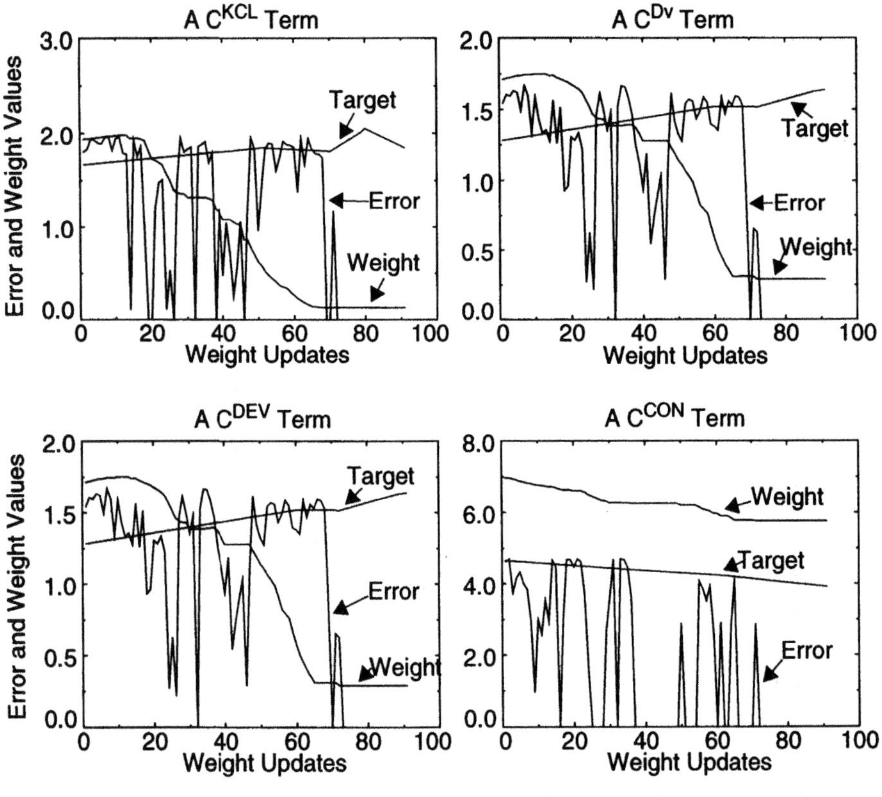

FIGURE 7.11 The Dynamic Weight Mechanism

error is larger than the target, the weight is increased, forcing the annealer to focus on that error and reduce it. This behavior is shown at the beginning of the annealing process for all but the last graph. On the other hand, if the error is below the target, the weight can be relaxed, allowing the annealer to focus on other terms in the cost function. This can be seen in all the graphs. In the C^{KCL} and C^{DV} graphs, the weights decrease significantly about halfway through the run. At this stage in the anneal, the C^{DEV} and C^{CON} terms are the annealer's main focus so the weights on the other terms are relaxed. (This can be seen by examining Figure 7.13, which was collected from the same run.) Thus, the dynamic weight control mechanism manipulates the cost term weights as the annealing process proceeds, forcing the errors to follow predetermined trajectories. As we have seen with the modified Lam cooling schedule, move selection mechanisms, and dynamic weights, inclusion of sophisticated automatic mechanisms can create an optimizer that is robust and does not require user manipulation of optimization constants for each new circuit problem.

7.5 The Relaxed-D.C. Formulation

In this section, we turn our attention from automatic control mechanisms to validation of the relaxed D.C. formulation. The ability of the annealer to examine circuit performance for circuits without first solving for the D.C. operating points is, of course, one of the fundamental concepts behind the ASTRX/OBLX framework. So, quantifying this process is essential to determining if this concept is effective in practice. The relaxed-D.C. formulation used in ASTRX/OBLX strives to ensure that the circuits are D.C. correct by placing a term in the cost function that penalizes differences of D.C. correctness, *i.e.*, C^{KCL} and C^{DV}. We have argued that early in the optimization process, exact D.C. bias points are not essential as the annealer can still get significant guidance from approximate D.C. bias points without the expensive complete solves.

In this section we look at experimental evidence to quantify this qualitative hypothesis. Figure 7.12 is a graph of average estimated voltage error across *all* the nodes in circuit *Simple OTA* as the optimization proceeds. Here, the average error hovers in the 100mV region. For errors of this magnitude, non-linear device models provide significant useful information about circuit performance. As the optimization nears completion, these errors are dramatically reduced until they are of the order found in the solutions generated by most detailed circuit simulators, and the associated errors in performance become insignificant. Thus, we conclude that the relaxed D.C. formulation provides enough information to guide the early stages of the annealing process while allowing convergence to an exact solution as the process completes.

7.6 The Annealing Process: Cost Function Evolution

The final aspect of the annealing process we discuss in this chapter is the pattern formed by the components of the cost function as the solution emerges. For the five types of terms that compose the cost function, Figure 7.13 shows the contribution of terms of that type to the overall cost as an annealing run proceeds. Recalling the cost term definitions from Section 4.4, here $C^{DC} = C^{KCL} + C^{DV}$, and C^{AWE} is typically 0 and not shown) The horizontal axis plots regular points in the annealing run where statistics of the cost function were sampled, and can reasonably be regarded as time. The data shown is typical and was collected for the circuit Simple OTA. The top graph is a composite showing all five of the traces. For clarity, the four graphs below show only a pair of terms. Each shows a different penalty term and includes the objective term. As the annealing run proceeds, the costs have a general downward trend, but it is not until after the halfway point that the terms shrink dramatically. Recall from Figure 7.8 that the temperature is reduced at about this same place, forc-

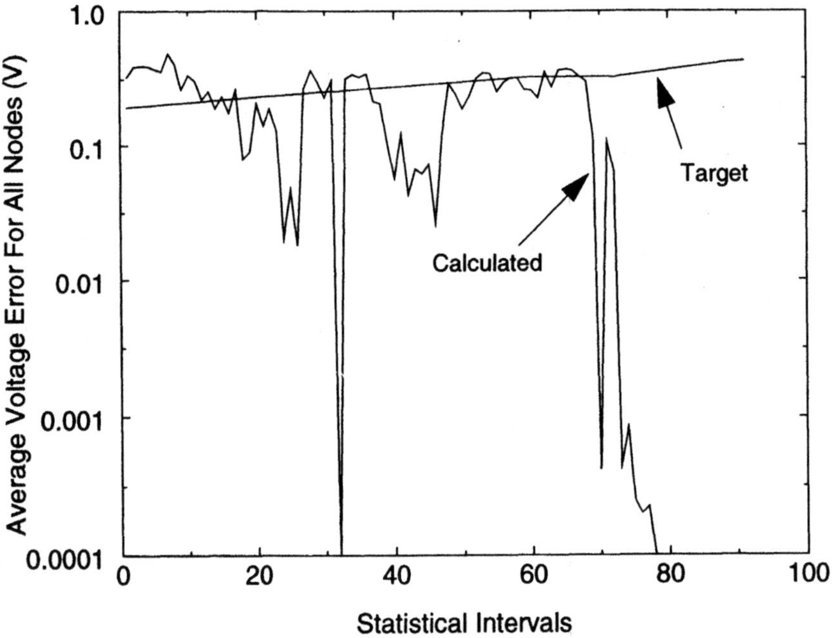

FIGURE 7.12 Evolution of Average Voltage Error During the Annealing Process

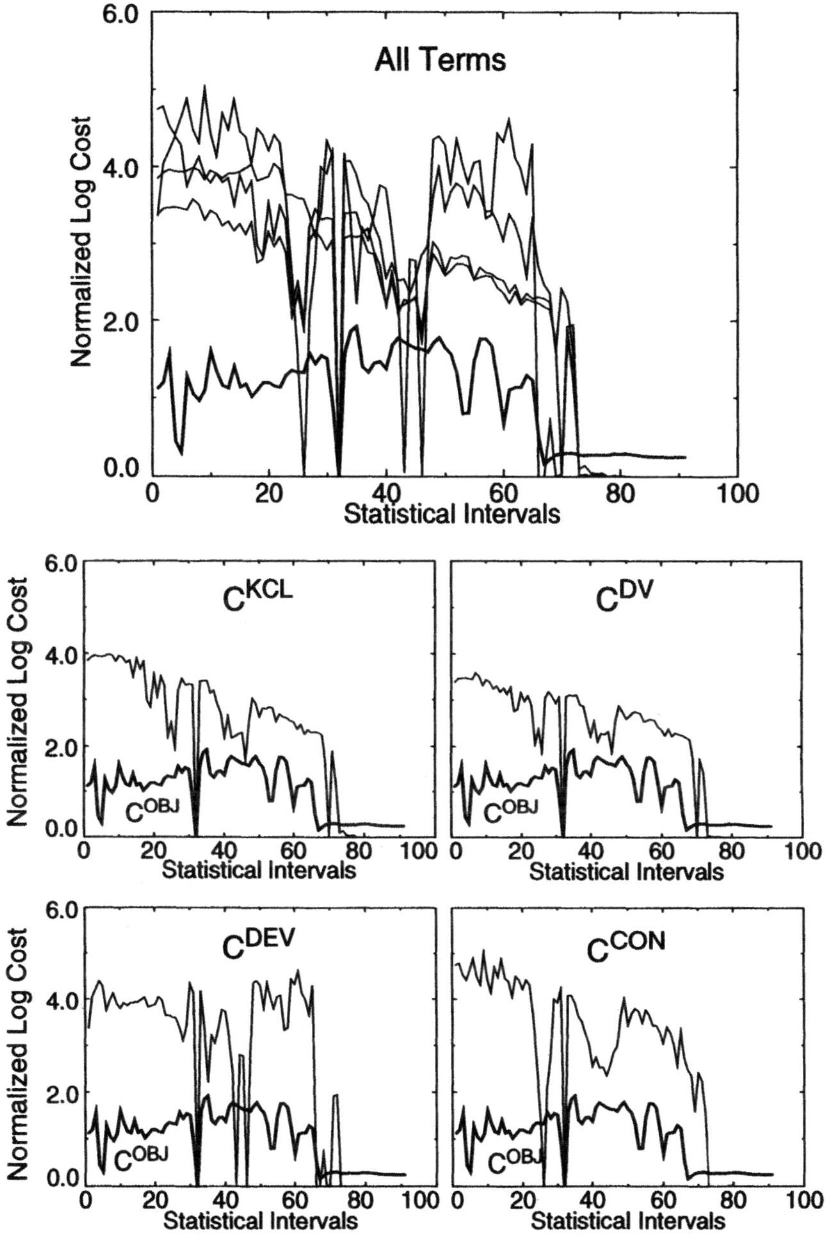

FIGURE 7.13 Components of the Cost Function During Annealing

ing the annealer to find a valley in the cost function and work towards its bottom. As expected, all the penalty terms are forced to zero, leaving only the objective term. It is interesting to note that the penalty terms disappear within a short space of time soon after the Lam schedule forces the accept ratio (and temperature) to decrease. In general, the device terms disappear first, indicating that all the transistors are in the correct region of operation. Next, the constraint terms go to zero, indicating that all the performance constraints are met. Finally, the D.C. terms go to zero, indicating that the D.C. operating point has converged. The graphs show that the annealing control mechanisms force the penalty terms in the cost function to converge to a feasible solution—leaving only the objective terms—soon after the accept ratio trajectory forces the system to cool towards its final solution.

7.7 Summary

In this chapter we presented experimental results that verify the effectiveness of the design of the annealing control mechanisms in ASTRX/OBLX. We presented a technique for tuning the annealer's control parameters by "optimizing the optimizer", then presented histograms that showed the effectiveness of the overall tuned annealer across several design problems. We then presented collected data that showed how some of the more important annealing control mechanisms behaved during typical annealing runs. And finally, we discussed the evolution of the annealing cost function representing a typical circuit design, both in terms of average voltage error across the nodes in the circuit and in terms of the types of components in the cost function.

This concludes our evaluation of the nominal synthesis formulation. We have shown how this formulation makes significant progress in three key metrics for synthesis tools—accuracy, automation, breadth. We now turn attention to the fourth metric, variation tolerance, and extend our formulation to include this final key aspect, completing our formulation of a practical analog circuit synthesis system.

CHAPTER 8

The Second Challenge: Handling Variations

Our work to this point has focused exclusively on solving the nominal synthesis problem for analog cell design. But, in order to create circuits that are practical, *i.e.*, useful in the real world, a synthesis system must create circuits that still behave correctly when there are operating point and manufacturing line variations. We refer to such a synthesis system as *variation-tolerant*. Extending our nominal synthesis approach to be variation-tolerant is the second challenge addressed in this book and the subject of the remaining chapters. In Section 1.3 we presented a simplified view of the mixed analog/digital design process, and described the role of variation-tolerant design in the cell design step of that process.

In this chapter, we begin with a detailed look at why nominal synthesis alone is not practical because of problems with manufacturing line and operating point variations. We then review relevant work from the area of parametric yield maximization. With manual designs, parametric yield maximization is used to minimize sensitivity to manufacturing line variations, so this work is an important area to draw from to create a variation-tolerant synthesis approach.

8.1 Problems with Variations in Nominal Synthesis

As validated in Chapter 6 and Chapter 7, the nominal synthesis strategy we described in Chapter 3, Chapter 4, and Chapter 5 is effective at solving the nominal parameter selection problem. Unfortunately, from the circuit or mixed-signal system designer's point of view, nominal synthesis is necessary, but not sufficient. What the designer wants is a cell that, after fabrication, can meet the performance goals in its operating environment. In this section we use two examples to show how the results of nominal analog synthesis systems can fall short of this goal.

8.1.1 Operating Range Failure

To highlight the limitation of nominal synthesis for the design of cells that can be used in a real world environment, we will revisit the comparison of a manual design and an automated design of the same cell. The cell, shown again in Figure 8.1, is a folded cascode amplifier [Nakamura 92] for which we have access to the complete circuit topology and device sizes originally obtained by an expert designer as well as the original process for which it was designed.

Recall that in Chapter 6 we used this circuit to compare nominal synthesis with ASTRX/OBLX to manual design. In that comparison, the manual design was simulated to obtain its performance. This same performance was given to ASTRX/OBLX as performance specifications, with one minor modification. The synthesis system was asked to synthesize a circuit that maximizes the unity gain frequency of this cir-

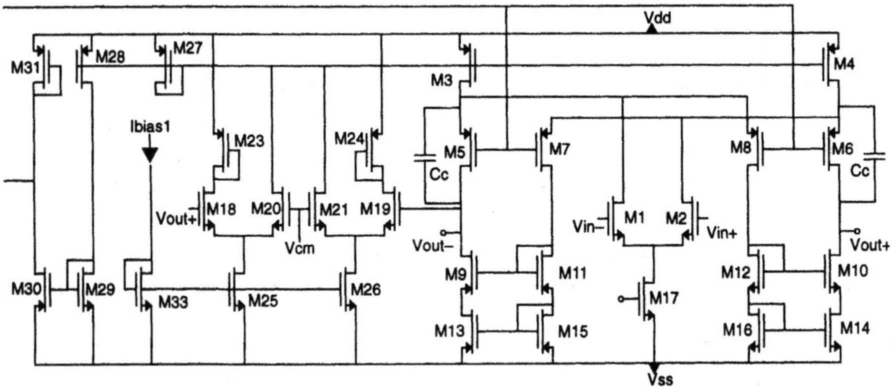

FIGURE 8.1 Novel Folded Cascode Operational Amplifier: Circuit Schematic

cuit while at the same time minimizing the area and maintaining the remaining performance specifications.

In Figure 8.2 we show the resulting D.C. gain performance of the manually designed operational amplifier and the one synthesized in Chapter 6 (marked OBLX). The synthesis was done at the nominal power supply level of $V_{dd} = 5.0V$, and the circuit definitely meets the design specification of $DCgain = 70dB$, at that nominal operating point. However, several devices are biased near the edge of saturation and a small decrease in power supply voltage dramatically decreases the gain. We can see the failing of nominal synthesis when we consider the operating environment for this cell. In its operating environment the cell may see a power supply operating voltage anywhere in the specified range $4.5 \leq V_{dd} \leq 5.5$. For almost half of this range, the circuit will not work as expected; this synthesized circuit is not practical because it will not meet performance targets under real world operating conditions.

We see further examples of this sensitivity to the operating point when we consider other circuit specifications as shown in Figure 8.3 and Figure 8.4. In a sense the optimization used by ASTRX/OBLX does its job too perfectly, thus leaving no room for changes in the operating point in almost all the performance specifications. A simple workaround to overcome this problem is to consider the performances at all the corners of the operating ranges. In this approach, the number of corners will grow to 2^m,

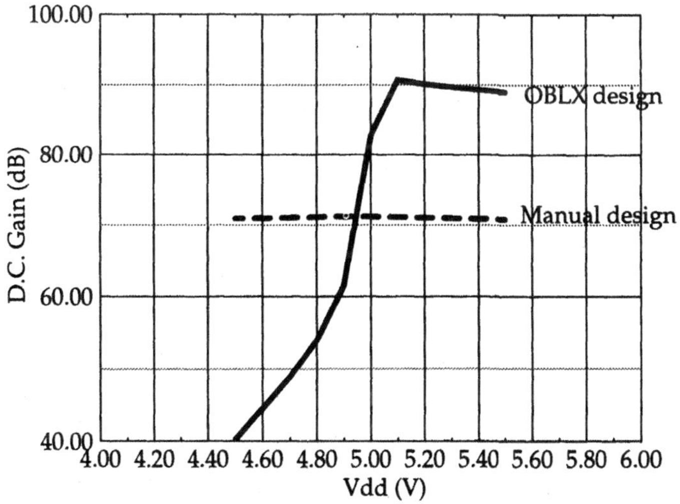

FIGURE 8.2 Variation of D.C. gain of Manual and Synthesized Design with Vdd

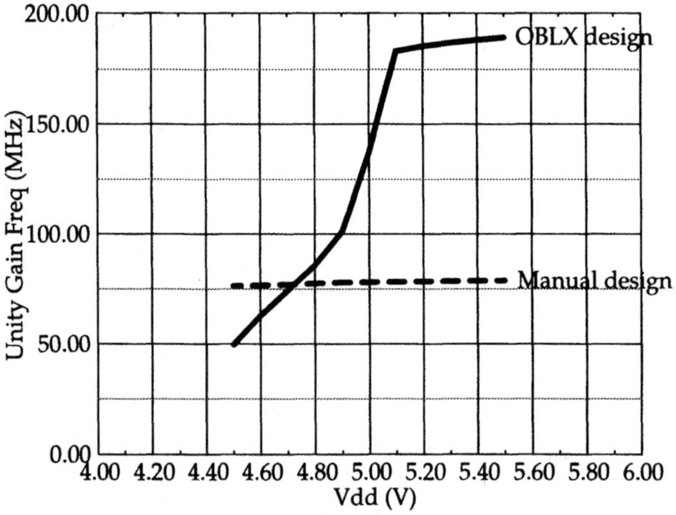

FIGURE 8.3 Variation of Unity Gain Frequency of Manual and Synthesized Design with Vdd

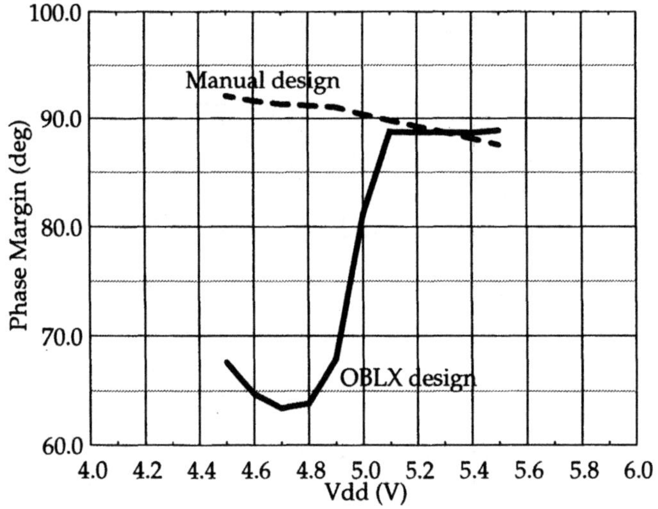

FIGURE 8.4 Variation of Phase Margin of Manual and Synthesized Design with Vdd

where m is the number of "worst-case" variables considered during the design. This exponential growth is problematic: it leads to excessively long run-times since the number of necessary performance evaluations grows exponentially.

An alternative is to use a limited number of corners which can be *a priori* specified by the designer as being the *critical* corners. This is similar to the approach often used in industry today: a fast process, a slow process, a high temperature point, a low temperature point, and so on. It is becoming increasingly difficult to predict these points when one does manual design. In the case of optimization-based automated synthesis, this is almost impossible to accomplish. There is no way the designer can know what path the optimization will take on its way to the final design. Almost by definition, an optimization technique will push the design in the direction it is least constrained. It would be almost impossible to pick the critical corners needed to completely constrain the optimization away from choosing a impractical solution, like the one picked in Figure 8.2, Figure 8.3 and Figure 8.4.

Moreover, since the worst case point is not necessarily at a corner, evaluation against operating range corners may not actually produce a design with adequate variation tolerance. As seen in Figure 8.4, the worst case point can occur within the range. Therefore we need a synthesis system that can determine these worst-case points dynamically "on the fly" as it evolves the circuit during the synthesis process.

8.1.2 Poor Yield Due to Manufacturing Variations

Previous approaches to design automation for cell-level circuits can be divided into two distinctly separate areas: analog circuit synthesis and yield optimization. The goal of analog circuit synthesis tools is to eliminate many of the circuit designer's most tedious tasks. The goal of yield optimization tools is to improve the yield of an already well-designed circuit prior to fabrication. While both sets of tools were aimed at helping the designer, they both solved just half the problem. The analog circuit synthesis tools often create designs that are at the edge of the performance space, whereas a good human designer using a familiar topology knows — at least approximately — how to over-design to ensure adequate initial yield. This implies that yield optimization tools will have more difficulty with nominally synthesized circuits than manually designed circuits.

We can see this when we make the same comparison between nominal synthesis and manual design as we did with operating range variations in the previous section. In Figure 8.5 we use the metric of post-design yield optimization to compare these two circuits. More precisely, we start off with two nominal designs: one manual and one synthesized by ASTRX/OBLX. We use yield optimization to improve the yield of both these circuits. Instead of the traditional yield optimization algorithms, we pursue an indirect method detailed in [Krishna 95]. This method uses a penalty function that penalizes designs that do not meet the desired circuit performances, while at the same time trying to improve the circuit's immunity to manufacturing line variations. At every iteration in the gradient-based approach, we determine the yield using a Monte Carlo analysis. This yield is plotted as a function of the iteration number in the graph in Figure 8.5. This graph shows that these algorithms work very well with the manual design, starting from an initial yield around 75% and taking the design to 100% yield. In comparison, the nominally synthesized circuit from ASTRX/OBLX *is not* a good starting point for the yield optimization. In this case the yield optimization is unable to affect the yield of the circuit. The nominal optimization in ASTRX/OBLX leads to a synthesis result that is trapped in a local minimum in terms of yield optimization. As a result, the yield optimizer cannot find the better minimum found using the manual design as a starting point. In fact, the yield optimizer cannot improve the yield of the

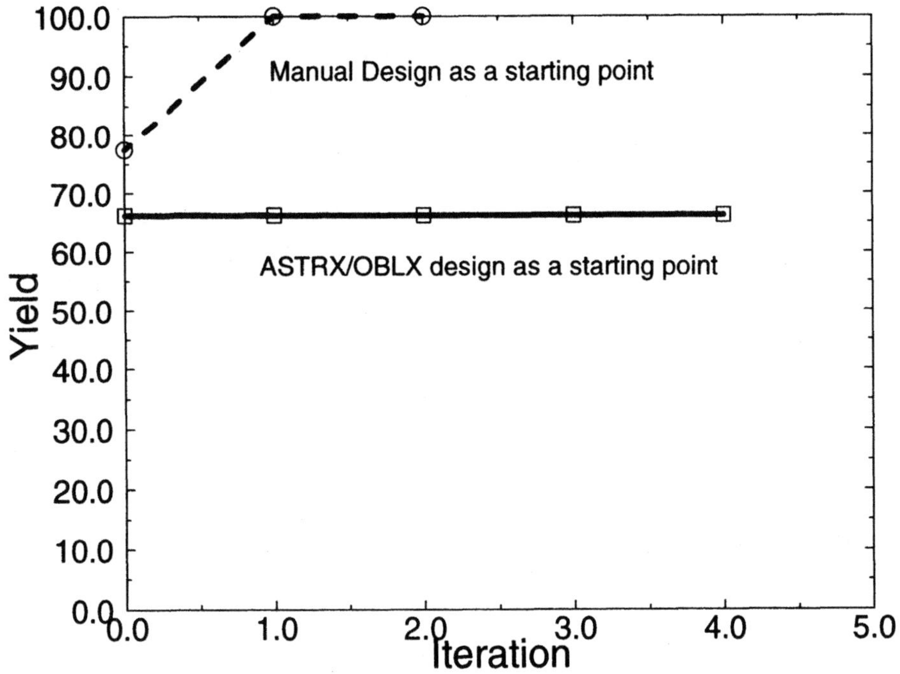

FIGURE 8.5 Post-Design Yield Optimization on Manual and Nominal Designs

design at all. The fact that the circuit is trapped in a poor local minimum is significant for all yield optimization approaches, since they tend to rely on gradient-based optimization just like the method in [Krishna 95]. Note that the difference between the manual design and the automated design starting points is simply the selection of the design parameters such as device sizes and bias values. The circuit topology, the circuit specifications, and the model of the manufacturing line variations were the same in both cases. This yet again highlights the local minima problem often seen with gradient-based optimization.

A designer intent on generating a working design knows of the variations in the manufacturing process and aims to reduce all sensitivities of the circuit performances to any variables not under her control to maximize the probability of a working cell at the end of the fabrication line. On the other hand, the nominal synthesis system is directed to meet the specifications *only* at the nominal point and, in effect, to ignore all variations. This difference is shown in Figure 8.6, which graphs two circuit performances, D.C. Gain (on the left y-axis) and Output Swing (on the right y-axis) with respect to a single manufacturing line variation (V_{FBP}, the pmos flat-band voltage as

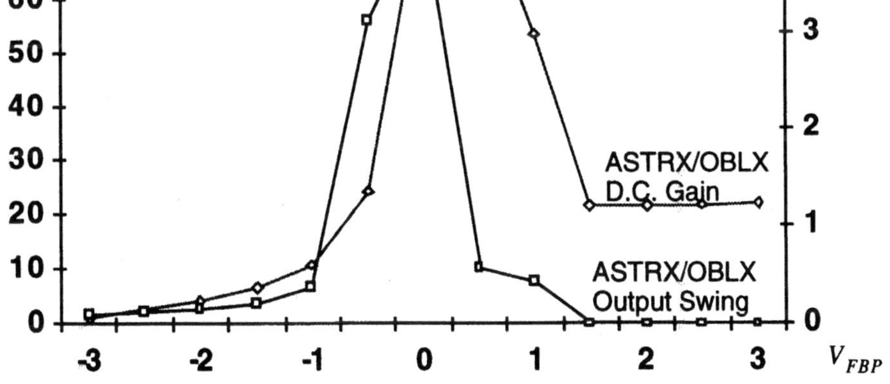

FIGURE 8.6 Effect of Manufacturing Variation on Circuit Performance

it ranges from -3σ to 3σ on the x-axis). The two circuit performances for the manual designs can be recognized by their insensitivity to the manufacturing line variation. This insensitivity arises from the designer's aim of selecting design parameters to meet specifications while simultaneously minimizing the sensitivity of the design's performance with respect to any variations.

In contrast, the nominal synthesis system leads to a design the meets its specifications only at the nominal point, with significant sensitivities to the manufacturing line variation. The result is a poor initial yield for the automatically generated initial design. In addition, the significant sensitivities with respect to the variations in the manufacturing line indicate that this design is at a local minimum with respect to the yield function. Any perturbations of the design parameters will either reduce or hardly affect the yield, so there is little scope for these perturbations to improve the yield. In fact, in our example for Figure 8.5 we ignored the Output Swing performance to aid the yield optimization with a large initial yield.

We believe that yield optimization tools can significantly improve the yields of good manual designs, but that an automatically synthesized circuit is often a bad starting point for gradient-based post-synthesis yield optimization tools. As we will demonstrate, device sizing is frequently driven by manufacturing or operating point variations. Synthesis based only on the nominal operating and manufacturing point can result in dramatically different device sizes/biasing, which may be incompatible with the assumptions of most yield optimization tools. For this reason, to create a practical synthesis system, we must extend our nominal synthesis approach to design for operating point and manufacturing line variations.

8.2 Parametric Yield Maximization

To date, in other work and in the first chapters of this book, the term *analog circuit synthesis* has meant nominal analog circuit synthesis. In the remainder of this book we extend this meaning to include operating point and manufacturing variations. Therefore we need to consider the previous work in the broad area of *design for manufacturability* to understand the limitations with respect to circuit synthesis.

The study of yield maximization began in the early 70's soon after nominal circuit optimization was introduced. These techniques developed mostly independently. In essence, nominal circuit optimization was promoted for designers who were interested in designing a single circuit, and parametric yield maximization was aimed at the design of circuits that would be produced in large numbers [Brayton 81].

The aim of yield maximization is to minimize the effects of IC manufacturing line fluctuations on IC performance. *Yield* is the name given to the term referring to the ratio of the number of working ICs to the number of manufactured ICs. There is a high correlation between this number and the profit generated by that particular IC, hence, yield maximization is an important component of circuit design. Yield loss in integrated circuits is generally due to two forms of manufacturing line disturbances [Maly 86]: *local* and *global*. Local disturbances, caused by particle defects in the fabrication line, manifest themselves as functional failures, *i.e.*, they change the basic circuit functionality. Global disturbance is the name given to phenomena that can be seen throughout the IC wafer, and can be caused by variations in temperatures, mask misalignments, and other individual steps of the fabrication process that might be incorrectly applied or controlled. These disturbances tend to cause parametric failures: the chip does "function," but fails to meet the parametric values of the performance specifications. In analog cells, due to their relatively small size and the continuous nature of their performance characteristics, the parametric yield loss mechanism overwhelms all yield loss due to particle defects, hence we concentrate on parametric yield maximization.

There have been two different areas of emphasis in parametric yield maximization: *process* and *circuit*. Process yield maximization refers to approaches that minimize the effects of the global disturbances by choosing the IC process controls (such as diffusion times and temperatures, *etc.*) suitably for a given circuit. Circuit yield maximization, on the other hand, takes a given process, and chooses the circuit designable variables to best maximize the expected yield in this given fabrication process. We limit discussion to circuit yield optimization since it shares the same variables as analog circuit synthesis.

Before we consider past research in parametric circuit yield maximization, it is important to define some notation related to this problem. Just as in circuit synthesis, there are parametric performance specifications. We can use these specifications to determine an *acceptability region*: the region of the space of all performances which meets the specifications for the design. While we define this in the performance space, this region can be mapped into other relevant spaces, including the space of designable variables for the circuit — which is the space we consider in circuit synthesis.

In circuit synthesis, we assume that the performances are functions of the designable variables and the nominal manufacturing and operating parameters. In this section we relax the nominal manufacturing parameters: the performances are now a function of a set of random variables which represents variation in the fabrication line. These random variables need to be modeled: in the yield optimization literature it is traditional to use a joint probability distribution function or *jpdf.*

We can build a definition of yield based on these two definitions (acceptability region and jpdf). Yield is the number of circuits that meet performance specifications after fabrication divided by the total number of circuits that have been fabricated. This corresponds to the probability that the random variables that model the manufacturing line variations belong to the circuit's acceptability region. This probability can be computed as the integral over the acceptability region of the jpdf, as shown in Equation 8.1,

$$Y(\underline{x}_0) = \int_{A_\xi(\underline{x}_0)} f_\xi(\underline{x}_0, \underline{\xi}) d\underline{\xi} \qquad \textbf{(8.1)}$$

where, $\underline{x}_0$ is the nominal value for the vector of statistical design variables in the circuit, $\underline{\xi}$ is the vector of statistical variables, $f_\xi(\underline{x}_0, \underline{\xi})$ is the joint probability distribution function of the disturbances at the given value for the design variables, and $A_\xi(\underline{x}_0)$ is the acceptability region. Note that we have not defined the space for the acceptability region. It could be the performance or parameter space or disturbance space [Feldmann 93], but it is simplest to consider the space for which we have the jpdf, $f_\xi(\underline{x}_0, \underline{\xi})$.

Finally, we have to maximize this yield, which means wrapping an optimization algorithm around the solution of the multi-dimensional integral in Equation 8.1 leading to:

$$max_{\underline{x}_0} \quad Y(\underline{x}_0) = \int_{A_\xi(\underline{x}_0)} f_\xi(\underline{x}_0, \underline{\xi}) d\underline{\xi} \qquad \textbf{(8.2)}$$

Therefore the four tasks in yield optimization are: determining the acceptability region, determining the jpdf, solving the yield integral, and optimizing this yield.

A *direct* yield optimization method is one that directly optimizes for yield. The yield integral can be computed via a Monte Carlo method. Monte Carlo methods however require many random samples, and can be costly since each sample usually involves a complete circuit simulation [Leung 74]. To overcome this, importance sampling can be used to reduce the number of Monte Carlo samples [Hammersley 64, Kalos 86].

There are two sets of commonly used approximations to avoid the computational complexity of yield optimization. The first set of approximations involve approximating the integration process. These methods are called *indirect methods*. In addition to reducing the computational cost needed for the optimization, they also have the benefit that they require little knowledge about the jpdf of the manufacturing line variations. An example indirect method is *design centering*. In this formulation the aim is to move the nominal—average—design so that it will be at the center of the acceptability region. This can be done either by inscribing the maximal norm body in the acceptability region [Director 77] or by pushing the acceptability region boundaries as far away as possible from the nominal point [Hocevar 88]. These methods simplify the yield optimization problem by assuming that either the acceptability region is fixed and the jpdf is a function of the circuit design variables, or that the jpdf is fixed and the acceptability region is a function of the circuit design variables. In a real design both the jpdf and the acceptability region are functions of the circuit design variables, which is why we use the explicit functions $f_{\xi}(\underline{x}_0, \underline{\xi})$ and $A_{\xi}(\underline{x}_0)$ in the expression for yield in Equation 8.1.

The second approximation applied to easing the computational complexity is related to approximations of the acceptability region. Since these methods explicitly formulate the acceptability region on the way to computing yield, they are called *explicit methods*. The earliest explicit methods were geometrical in nature [Director 77], however recent methods use analytical approximations of the acceptability region with first or second order polynomials.

Analytical approximations of the circuit performances can also be made. This reduces the computational complexity of the circuit simulation needed in *implicit methods* (those that compute yield without computing an acceptability region). These analytical approximations can then be surrounded by a general non-linear programming algorithm for optimization. For traditional non-linear programming algorithms this requires an efficient and accurate evaluation of the yield gradient in addition to the yield [Feldmann 93].

Of all the approaches presented, [Feldmann 93] is most promising in its ability to manage both jpdfs and acceptability regions that vary with designable variables. Feldman's boundary integral formulation is a very efficient formulation since it provides a uniform mechanism to compute the circuit yield and its yield gradient.

There are two major limitations of prior yield maximization approaches. First, while there have been some attempts to link operating point variations with yield maximization strategies [Bandler 78, Brayton 80], these two problems are usually separated. For a circuit to be deemed manufacturable, both these issues need to be handled. Second, to reduce the number of yield and yield gradient evaluations, yield maximization methods tend to use gradient-based non-linear programming algorithms for optimization. This limits their usefulness to circuits that have been designed for non-zero initial yield with non-zero initial gradients. This is not a real limitation for human designers, who tend to incorporate a simple but realistic view of the yield detractors when creating their *nominal* design. However, as shown by experiments such as Figure 8.5, when we tie yield optimization approaches to *automatic* nominal synthesis approaches we see problems arising from severe incompatibilities of philosophy.

A third problem that has limited the use of statistical IC design techniques amongst designers is the lack of models that can be used during the design. Unlike nominal design, where it is easy to get a characterization of the process for which the circuit is being designed for, creating a statistical model of the fabrication line has proved to be difficult. Recent work in statistical parameter extraction [Krishna 95a] is however showing promise in providing both a suitability metric of a device model, and extracting a designer-friendly statistical model of the manufacturing process.

8.3 Summary

We have seen that although the nominal synthesis system presented in the first chapters of this book is very accurate, highly automated, and broadly applicable, it is still not practical because it is not tolerant of variations in the operating environment and manufacturing line. With manually design circuits, yield optimization techniques can be used to alleviate this problem, and we reviewed existing techniques to do so. Unfortunately, the results of nominal synthesis systems like ASTRX/OBLX are potentially incompatible with post-synthesis yield optimization. To overcome these limitations to synthesis is the second major challenge that must be overcome to create a practical synthesis system. To meet this challenge, we create a new synthesis formulation that unifies the various senses of the parameter selection problem: nominal performance, operating point ranges, and statistical manufacturing variations. The following chapters describe this formulation in detail, present an optimization technique to design circuits with this formulation, and validate our work by presenting circuit design results.

CHAPTER 9

A Unified Formulation

In this chapter we present a new formulation of the circuit synthesis problem. This formulation overcomes the limitations of the nominal synthesis formulations that we described in the previous chapter. We begin this chapter with a description of the goals of our formulation. We then introduce the concept of the range variable and state the mathematical formulation of the variation tolerant circuit synthesis problem. Finally we present an example to clarify the components of this formulation.

9.1 Problem Formulation Goals

We have defined the goals for analog cell synthesis in Section 1.4. We also defined four metrics to evaluate achievement of these goals: accuracy, automation, breadth, and variation tolerance. Our nominal synthesis formulation, as detailed in the first part of this book, does well with accuracy, automation, and breadth, but falls short with variation tolerance. In this chapter we describe our extensions of our nominal synthesis formulation to handle variation tolerance. We begin by formulating three additional goals for our unified formulation intended specifically to address the variation tolerance problem.

The aim of modelling the analog circuit design process is to create a representation of the analog circuit design problem keeping in mind the three following features:

1. **Completeness.** The model should describe the complete design problem. Analog design is often given the status of *a black art* because expert design experience has a large affect on quality of design. An expert analog designer is successful in solving a design problem by appropriately deciding what factors can be assumed away, and what factors need to be taken into account. An automated system needs to start with a complete model, and needs to make its own decisions of what to ignore during the synthesis process. Earlier attempts at creating models of the design problem have traditionally modelled only part of the design problem; implementations of these incomplete models tend to generate designs that require a generous amount of manual tweaking of the synthesized design.
2. **Unified Formulation.** As we saw in Section 8.1, nominal synthesis needs to be extended to include operating point variations and fabrication line variations. Both these issues also affect the sizing and biasing decisions made during circuit design, hence we will use a single *unified formulation* to determine the circuit parameters. By a unified formulation we mean that we will treat nominal synthesis, operating point variations and fabrication line variations in identically the same manner. This will entail some simplifications of the problem; we shall discuss these in detail in Section 9.4. This can be contrasted with the division of the problem into several sub-problems that are solved sequentially, as implied by the flow in Figure 1.2 on page 5. Such a divide-and-conquer scheme is applicable when division is possible. In the case of variation tolerant synthesis the optimization variables for nominal sizing and biasing and those for controlling variation tolerance are identically the same variables, thus we need to use a unified formulation.
3. **Usable by a computer.** This practical goal is important because of its relation to the two most important goals for ASICs: reduced time-to-market and reduced design costs. Creating a model that requires manual intervention will not attain these goals. The ultimate aim of our unified formulation is to create an analog cell synthesis system which will use a computer to design cells on demand for an ASIC designer. With probability distribution functions and yield integrals, it is much more likely than with nominal synthesis to create a formulation that cannot be coded in a practical fashion. Thus any mathematical model of the design problem needs to be able to be coded in a synthesis tool.

As with our nominal synthesis formulation, achieving our goals requires some trade-offs. We are still willing to make the same trade-offs described in Section 3.1—longer run times and no topology selection. However, unlike for our nominal synthesis formulation, we are no longer willing to concede variation tolerance. The result will be a greater investment of run time, but it will be well spent because the investment will yield practical circuit designs.

9.2 Strategy

Our three goals for our variation tolerant formulation are addressed by the addition of key concepts to our existing nominal synthesis formulation. Recall from Chapter 3 that our nominal synthesis formulation relies on the following five key ideas:

1. A non-linear constrained *optimization formulation* of the circuit synthesis problem that we map to an unconstrained optimization algorithm and solve numerically.
2. A *relaxed-D.C. formulation*, that avoids a CPU intensive D.C. operating point solution after each step in the optimization.
3. *Simulated annealing* [Kirkpatrick 83] as the optimization engine to provide robustness and the potential for global optimization in the face of many local minima.
4. *Asymptotic Waveform Evaluation* (AWE) [Raghavan 93, Pillage 90] augmented with some simple, automatically generated analytical equations to predict circuit performance.
5. A compiled database of industrial quality, non-linear device models, called *encapsulated device evaluators*, that provides the accuracy of a general-purpose simulator while making the synthesis tool completely independent of low-level device modelling concerns.

To achieve the additional goal of variation tolerance, we must rely on three additional key ideas:

1. Introduction of *range variables* to model supra-nominal variations. These variations include the circuit environment operating point fluctuations and the variations in the fabrication line.
2. Formulation as a *Non-linear Infinite Programming* problem to completely model the circuit design problem as a circuit synthesis problem.
3. Solution of this formulation with a sequence of two alternating optimization problems: a *nominal synthesis* problem solved using the techniques described in Chapter 3, Chapter 4, and Chapter 5; and a *worst-case optimization* problem solved using annealing.

The first two points relate to modelling the design problem, and we consider them in the balance of this chapter. We have already detailed our nominal synthesis solution strategy in Chapter 3, Chapter 4, and Chapter 5. The worst-case optimization problem will be discussed in Chapter 10.

9.3 Notation and Formulation

We have already seen the nominal optimization formulation of a fixed-topology, parameter selection problem in Chapter 3. We described our formulation using definitions in Section 2.3.1. In this section we recall these definitions, extend them to simplify the presentation of our unified formulation, and finally present the unified formulation.

- Recall that $\underline{x}$ is the vector of n independent variables. These are the state variables manipulated by the annealer. For simplicity of explanation, we neglect the fact that the relaxed D.C. formulation requires the introduction of additional voltage variables.
- Recall that $\underline{f}(\underline{x})$ is the set of k objective functions.
- Recall that $\underline{g}(\underline{x}) \le 0$ is the set of l inequality constraint functions.

Sometimes a specification is made as a function of an independent parameter (*e.g.*, the gain should be greater than 60dB for all values of V_{dd} in the range $4.5 \le V_{dd} \le 5.5$). We need to introduce an operating point *range variable* to handle this constraint, which we denote as $\underline{g}(\underline{x}, \underline{\xi}) \le 0,\ \forall \underline{\xi} \in \Xi$, where $\underline{\xi}$ is the vector of operating point range variables, and Ξ is the set of values over which the operating point is expected to vary. In the gain example, the vector $\underline{\xi}$ is simply a scalar V_{dd}, and the set Ξ is the range $4.5 \le V_{dd} \le 5.5$, thus the constraint becomes

$$60dB - gain(\underline{x}, V_{dd}) \le 0 \qquad \forall V_{dd} \in [4.5V, 5.5V] \tag{9.1}$$

Examples of common operating point parameters include time, frequency, temperature and bias voltage.

Fluctuations inherent to a manufacturing process affect circuit operation, and need to be modelled. In this book, we choose an identical formulation for the operating point and manufacturing variations; so the constraints are again of the form of $\underline{g}(\underline{x}, \underline{\xi}) \le 0,\ \forall \underline{\xi} \in \Xi$. Unlike the operating point case, where we formulate the range variable in a rather natural way, for statistical manufacturing line variations, we introduce a significant simplification. The approach we use is to consider the region in the space of manufacturing line variations which is *sufficiently probable*. Therefore, instead of considering a joint-probability distribution function model of the statistical variations, we *simplify* the variations into a region in the space of manufacturing variations, which contains a *significant* mass of the probability distribution. Then we rewrite the circuit performance constraints such that the circuit will meet its performance specifications *for all* values in this sufficiently probable space. Hence both the operating point variations and the manufacturing variations involve the same type of variable, a *range variable*. We extend our set of definitions (See "Formulating Analog Design As A Numerical Optimization Problem" on page 28) to include range variables as follows:

Def 4: Let $\underline{\xi}$ be vector of p range variables. These variables include all the circuit environment operating point variables, and the random variables modelling the manufacturing line variations. The set of points in the set Ξ, of the $\Re^p$ range variable space defines the space in which we are interested. These points become a part of a *for all* constraint $\forall\underline{\xi} \in \Xi$ that is attached to at least one constraint or objective function. However, any and all constraints and objectives can have such a *for all* constraint. These variables are distinct from the design variables, $\underline{x}$, which the optimization is trying to determine.

With the introduction of range variables we need to redefine all the functions to be functions of this vector of variables. Therefore, $\underline{f}(\underline{x}, \underline{\xi})\ \forall\underline{\xi} \in \Xi$ is the new set of objectives and $\underline{g}(\underline{x}, \underline{\xi}) \le 0 \quad \forall\underline{\xi} \in \Xi$ is the new set of inequality constraints.

9.3.1 Complete Formulation

Now that the notation for the variables and the equations are defined, we can write the complete formulation as in Equation 9.2.

$$\left.\begin{array}{ll} min_{\underline{x}} & \sum_{i=1}^{j} w_i \cdot f_i(\underline{x}, \underline{\xi})\ \forall\underline{\xi} \in \Xi \\ \text{subject to} & \\ & \underline{g}(\underline{x}, \underline{\xi}) \le 0 \quad \forall\underline{\xi} \in \Xi \\ & \underline{x} \in X_P \end{array}\right\} \quad NLIP \qquad (9.2)$$

This equation is very similar to the optimization formulation shown in Section 2.3.1 on page 28. The major difference is the *for all* terms added to both the objective and the constraints. A more minor change is the addition of $\underline{x} \in X_P$ which formalizes the definition of the allowable ranges for the independent variables, *i.e.*, $x_i \in [x_i^{min}, x_i^{max}]\ \forall x_i \in \underline{x}$ introduced in Section 4.2 on page 84. This formulation is called a *Non-Linear Infinite Program* due to the infinite set of objectives and constraints.

It is important not to confuse the range variables $\underline{\xi}$ with the circuit design variables $\underline{x}$. At the end of the optimization we want one set of values for the vector of circuit design variables, and we wish that this set of values lies in the set defined by X_P. In contrast, we want the objectives and the constraints measured at this optimum $\underline{x}^*$ to be at their minima and less than zero respectively, *for all* the possible values of the vector of range variables in its set Ξ. If we mistakenly treat a range variable as a design variable, we can expect the following scenario: the optimum design occurs for a particular value of the range variable, say, $V_{dd} = 4.87V$. But we are not interested in a *single* value of the range variable for which the optimum occurs; rather we want to use the range variable to guide *where* we should evaluate the objective and constraint functions during the optimization to ensure that the *for all* constraint is met.

9.4 Example

Our mathematical programming formulation of the complete synthesis problem is best motivated using a simple example. The example traces both the elements that a typical circuit designer worries about in terms of the design problem itself, as well as the issues that she needs to worry about to describe completely the design problem to a computer.

Let us assume a designer wishes to design an operational transconductance amplifier like the one shown in Figure 9.1 for the specifications shown in Table 9.1. This is an extension of the example used in Section 3.4.1, except we now consider a single-ended operational amplifier where variations will affect the input offset voltage. Since most of this example is similar to the example in Section 3.4.1, we will proceed quickly through the parts that are equivalent in both the nominal and the variation tolerant synthesis formulation. We will first discuss the circuit specifications and then the manufacturing line and operating environment abstractions.

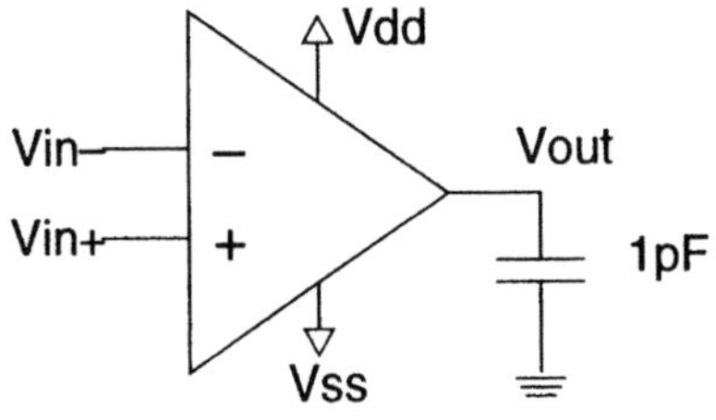

FIGURE 9.1 Design Example: An Operational Transconductance Amplifier

Assume that the circuit topology shown in Figure 9.2 has been chosen. The remaining degrees of freedom involve the device parameters, such as MOS geometries, for the circuit. In this example, the sizes of the current mirror devices are pre-determined to create a simpler example. The only other degrees of freedom are the width and length

TABLE 9.1 Design Example: Performance Specifications

Attribute	Specification
A_{dc}	maximize $\forall V_{dd} \in [4.5,5.5]$
UGF	$\geq$ 1MHz $\forall V_{dd} \in [4.5,5.5]$
PM	$\geq$ 60 degrees $\forall V_{dd} \in [4.5,5.5]$
SR	$\geq$ 1V/μs $\forall V_{dd} \in [4.5,5.5]$
OV	$\geq$ 2mV $\forall V_{dd} \in [4.5,5.5]$

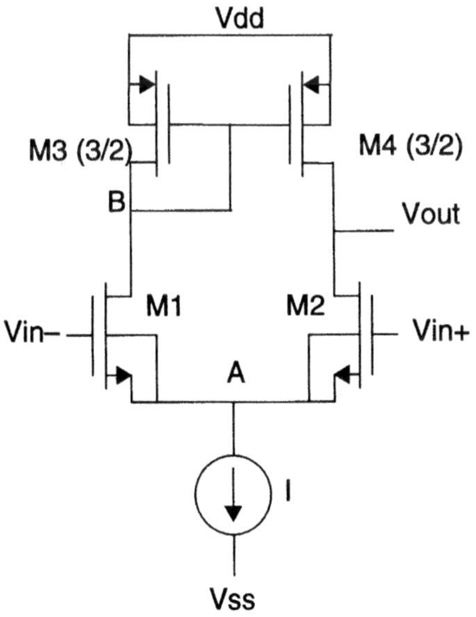

FIGURE 9.2 Circuit Schematic showing Sizing and Biasing Degrees of Freedom

of the differential pair devices (M1 and M2) and the current in the tail current source (I). Therefore the degrees of freedom in sizing and biasing this circuit are:

$$\underline{x} = \{W, L, I\} \tag{9.3}$$

As with our nominal synthesis example, even though two transistors are being sized in this differential pair, they are nominally expected to be of identical size, so only a single set of sizing variables, W and L, are necessary. Thus we can specify our netlist as:

```
Amplifier circuit {
        M1    width=W, length=L     nodes=(B,Vin–,A,A)
        M2    width=W, length=L     nodes=(Vout,Vin+,A,A)
        M3    width=3, length=2     nodes=(B,B,Vdd,Vdd)
        M4    width=3, length=2     nodes=(Vout,B,Vdd,Vdd)
        I     value=I               nodes=(A,Vss)
}
```

9.4.1 Circuit Specifications

As we discussed in Section 3.4.1, a circuit specification is composed of two parts: the parametric value of the specification, and a means of measuring the specification, often with a test jig. From Table 9.1, the specifications of interest are:

- A_{dc}, the D.C. voltage gain when the operational transconductance amplifier is loaded by a 1pF capacitance;
- UGF, the unity gain frequency;
- PM, the phase margin;
- SR, the slew rate;
- and OV, the input-referred offset voltage.

In Section 3.4.1 we saw how to handle similar specifications, with the exception of OV, the input referred offset voltage, so in this section we merely summarize the required information for the other specifications. To measure A_{dc}, UGF, and PM, we can use the test circuit shown in Figure 9.3. The D.C. gain, A_{dc}, is measured as the ratio (Vout/Vin) when a D.C. input, Vin is applied. Measuring UGF requires increasing the frequency of Vin until (Vout/Vin) is 1. Phase Margin, PM, can be computed from the phase measurement at this frequency of unity gain.

Example

The new specification in this example is the input referred offset voltage, OV, a practical concern when designing MOS op-amps. This is defined as the input-referred offset voltage $v_{in,\,off}$ required to restore the output voltage v_{out} to zero. The designer can zero this easily by choosing a fully differential topology as we did in Section 3.4.1, but when other design concerns dictate the use of a single-ended design, systematic offset voltage must be controlled with proper dimensions and/or bias conditions. Even though this is a non-linear specification, it can be specified with a test circuit to ASTRX since it does not involve any transient analysis. Figure 9.4 shows the test jig and a zero-valued current source that can be used to measure $v_{in,\,off}$ from this circuit. Internally, OBLX performs D.C. biasing during its optimization, and the input-referred offset can be measured from this analysis. We can write this test-jig as:

$v_{in,\,off}$ test jig {

Use the amplifier circuit

Vdd	value=[4.5V,5.5V]	nodes(Vdd, gnd)
Vss	value=0.0V	nodes(Vss, gnd)
Iin	value=0A	nodes(Vin+,gnd)
Vf	value=0V	nodes(Vin-,Vout)

calculate $v_{in,\,off}$ as V(Iin)

}

and the resulting constraint is:

$$v_{in,\,off}(\underline{x}) \leq 2mV \tag{9.4}$$

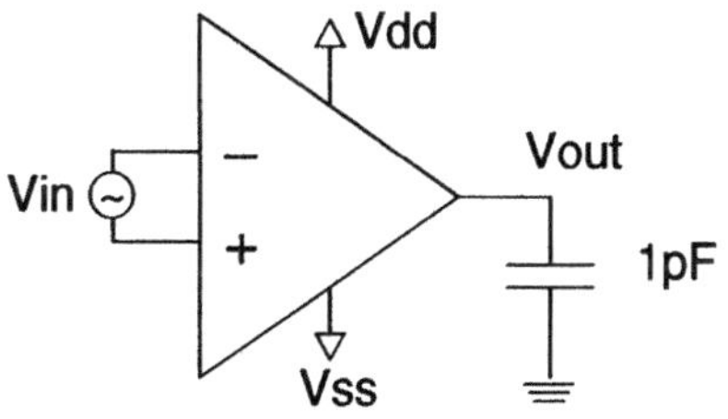

FIGURE 9.3 Simulation Test-jig for Measuring A_{dc}, UGF and PM

As with the other circuit specifications, the input offset voltage in the nominal design case is a function of the design variables $\underline{x}$, with the specification value obtained from Table 9.1. We can re-write this into the canonical form

$$v_{in,\,off}(\underline{x}) - 2mV \leq 0 \tag{9.5}$$

The offset voltage is the only new specification added to the example of Section 3.4.1, and it is essential because we are using a non-differential operational amplifier topology. As we shall see in the next section, this is a good test for our new unified formulation because the total input referred offset has both a systematic component (due to operating point variations) and a random component (due to manufacturing variations).

the systematic offset requires consideration of manufacturing line variations.

9.4.2 Operating Point Variations

We will now look at the non-nominal specifications. The first class of non-nominal specifications we consider is *operating point variations*. A circuit after fabrication will be operated in an environment that can vary around the nominal operating point. This environment includes parameters such as the V_{dd} power supply voltage, the operating temperature, and so on. We specify this variation by providing a range of allowable values for this parameter. This is called an *operating point range*.

In the case of our design example, there is a single operating point variation, V_{dd}. The operating point range is specified for each performance specification in

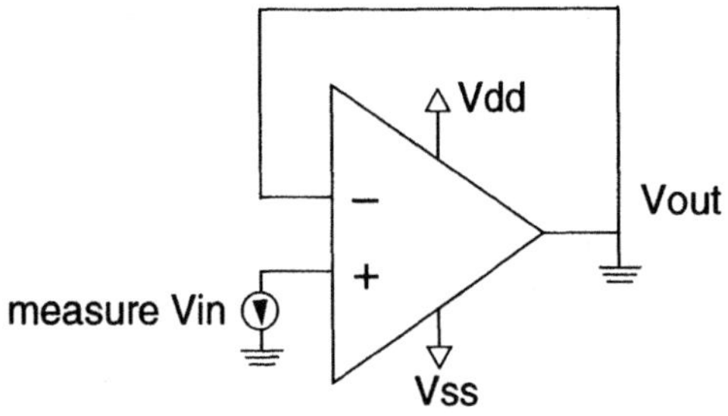

FIGURE 9.4 Test-jig for Offset Voltage Computation

Table 9.1. Each specification has to be met in the constraint given by $\forall V_{dd} \in [4.5,5.5]$. This means that each specification has to be met *for all* the possible values of V_{dd} in the range [4.5, 5.5]. Recall that each performance specification becomes a single constraint at the nominal point in the range. Now, with the introduction of the operating point range, we have an *infinite* number of constraints for each performance specification. This is because the performance constraint has to be met for *every* infinitesimal value of the range variable in the range. All our performance constraints will look exactly the same as before, with the additional *for all* term. For example, the unity gain frequency constraint now looks like:

$$1MHz - UGF(\underline{x}, V_{dd}) \leq 0 \ , \quad \forall V_{dd} \in [4.5,5.5] \tag{9.6}$$

9.4.3 Variations Related to Manufacturing Fluctuations

Statistical variations in the physical controls in the fabrication line lead to yield loss in integrated circuits. This is the last issue a designer has to consider in completing the model of the design problem. As seen in Section 8.2, yield maximization techniques can be used to determine the best design for a given set of specifications and a joint probability distribution of the fluctuations in the fabrication line. Our unified formulation attacks the problem of manufacturing line fluctuations, but in a different way than the yield maximization approach.

In our example circuit design problem, the manufacturing variation has yet to be modelled. Without such a model, we would be designing for the systematic component of the offset voltage, ignoring the often large random component. Unlike the standard test-jigs for nominal performance measurements, or the standard ranges for operating points, there is no standard model for manufacturing line variations. While there is significant current work in this area, a commonly used modeling approach is that the variations in a fabrication line can be modeled as a joint probability distribution function.

It has been shown that a relatively few device model variables can capture the bulk of the variations in a CMOS manufacturing line [Cox 85, Krishna 95a], with the most significant ones being threshold voltage, oxide thickness and the length and width lithographic variations. In practice, the oxide thickness variation is very tightly controlled in modern processes, so we will ignore it for our example. In addition, for analog cells, the devices widths are much larger than the lengths. Therefore the circuit performances will be more sensitive to the length variations than the width, so we will ignore the width variation for this example. As a result, we consider two sources of variation, intra-chip threshold voltage variation, ΔV_T and intra-chip length reduction variation, ΔL. We do not assume any statistical independence between these sources, however, we assume that all other parameters do not vary, and can be considered to remain at their nominal values. This simplification to two variations has been chosen so we can describe the jpdf graphically in the remaining discussion.

ΔV_T is the intra-chip difference in the threshold voltage, *i.e.*, the difference in the threshold voltage between device M1 and M2, given that they *should* be identical devices. A similar difference can occur for devices M3 and M4, but again for simplicity in this introductory example, we assume that there is no such difference for these current mirror devices.

ΔL is the intra-chip difference in the lengths of the M1 and M2 devices. Again we assume that there is no difference in the actual lengths of M3 and M4, *i.e.*, they remain at their nominal values.

We now need to develop a model of the fabrication line to determine the ranges for the constraints on ΔV_T and ΔL. While the simple model is to use the wafer acceptance specifications provided by the process engineers, we will consider a joint-probability distribution function: $p(\Delta V_T, \Delta L)$. This will help in comparing and contrasting with the approaches previously used for yield optimization. This jpdf can be determined using measurements from test-circuits fabricated in the manufacturing line or via statistical process and device simulation [Nassif 84]. We can draw the contours of $p(\Delta V_T, \Delta L)$ as shown in Figure 9.5 with the 3σ contour highlighted. In our model of the fabrication line variations, we will treat this contour in exactly the same way as a range variable. We therefore specify that the circuit being designed should work for *every* possible value of ΔV_T and ΔL within the 3σ contour. More exactly, we are constraining the yield to be at least 3σ.

Stepping back from the example to consider this general approach, we see that this is *not* a statistical IC design formulation of the synthesis problem. Instead it is a unified formulation that converts the problem of manufacturing line variations into the same numerical form as the operating range variations. We do this by converting the statistical distributions of the fluctuations in the manufacturing line into a range variable which spans a *sufficiently probable* range of these statistical distributions.

Since all the circuit performances are functions of these manufacturing line variations, we have to include them in the new functions. Thus, for example, the D.C. gain function changes from $A_{dc}(\underline{x}, V_{dd})$ to $A_{dc}(\underline{x}, \Delta V_T, \Delta L, V_{dd})$.

Our optimization problem is now complete, and looks like

$$\min_{\underline{x}} \quad A_{dc}(\underline{x}, \Delta V_T, \Delta L, V_{dd}) \quad \forall V_{dd} \in [4.5, 5.5] \ , \ \forall \Delta V_T, \Delta L \in 3\sigma$$

subject to

$$\begin{aligned} UGF(\underline{x}, \Delta V_T, \Delta L, V_{dd}) &\geq 1MHz \ , \ \forall V_{dd} \in [4.5, 5.5] \ , \ \forall \Delta V_T, \Delta L \in 3\sigma \\ PM(\underline{x}, \Delta V_T, \Delta L, V_{dd}) &\geq 60 degrees \ , \ \forall V_{dd} \in [4.5, 5.5] \ , \ \forall \Delta V_T, \Delta L \in 3\sigma \\ SR(\underline{x}, \Delta V_T, \Delta L, V_{dd}) &\geq 1V/\mu s \ , \ \forall V_{dd} \in [4.5, 5.5] \ , \ \forall \Delta V_T, \Delta L \in 3\sigma \\ v_{in,off}(\underline{x}, \Delta V_T, \Delta L, V_{dd}) &\leq 2mV \ , \ \forall V_{dd} \in [4.5, 5.5] \ , \ \forall \Delta V_T, \Delta L \in 3\sigma \end{aligned} \tag{9.7}$$

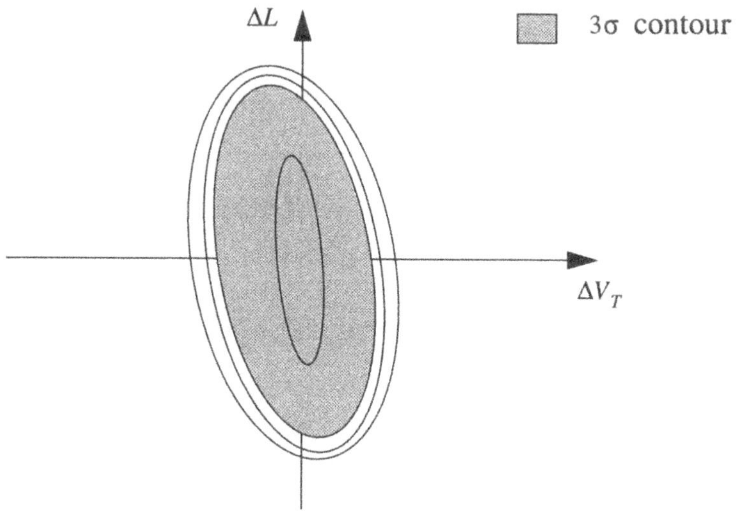

FIGURE 9.5 Contours of jpdf of Fabrication Line Model

which can be written in the canonical form to make it look like Equation 9.2:

$$\begin{array}{c} \min_{\underline{x}} \quad A_{dc}(\underline{x}, \Delta V_T, \Delta L, V_{dd}) \quad \forall V_{dd} \in [4.5{,}5.5] \ , \ \forall \Delta V_T, \Delta L \in 3\sigma \\ \text{subject to} \\ 1MHz - UGF(\underline{x}, \Delta V_T, \Delta L, V_{dd}) \le 0 \ , \ \forall V_{dd} \in [4.5{,}5.5] \ , \ \forall \Delta V_T, \Delta L \in 3\sigma \\ 60degrees - PM(\underline{x}, \Delta V_T, \Delta L, V_{dd}) \le 0 \ , \ \forall V_{dd} \in [4.5{,}5.5] \ , \ \forall \Delta V_T, \Delta L \in 3\sigma \\ 1V/\mu s - SR(\underline{x}, \Delta V_T, \Delta L, V_{dd}) \le 0 \ , \ \forall V_{dd} \in [4.5{,}5.5] \ , \ \forall \Delta V_T, \Delta L \in 3\sigma \\ v_{in,\,off}(\underline{x}, \Delta V_T, \Delta L, V_{dd}) - 2mV \le 0 \ , \ \forall V_{dd} \in [4.5{,}5.5] \ , \ \forall \Delta V_T, \Delta L \in 3\sigma \end{array} \tag{9.8}$$

The input offset constraint now has both a systematic component (independent of ΔV_T and ΔL) and a random component (dependent on ΔV_T and ΔL).

9.5 Comparison with Previous Formulations

Equation 9.2 contains one ambiguity in its objective that deserves further clarification. In the case of manufacturing range variables in the vector $\underline{\xi}$, there is a statistical distribution within the range with some areas of the range being more probable than the rest. Now, unlike the classical yield optimization, we have a truncated jpdf; but we can still integrate the jpdf leading to the following problem

$$\begin{array}{l} \min_{\underline{x}} \quad \sum_{i=1}^{j} w_i \cdot \left\{ \int_{\underline{\xi} \in \Xi} f(\underline{x}, \underline{\xi})\, p(\underline{\xi})\, d\underline{\xi} \right\} \\ \text{subject to} \\ \qquad g(\underline{x}, \underline{\xi}) \le 0 \quad \forall \underline{\xi} \in \Xi \\ \qquad \underline{x} \in X_P \end{array} \tag{9.9}$$

which is similar to the approach used in [Grossmann 91]. The only difference between Equation 9.2 and Equation 9.9 is the objective. Due to the statistical distribution in the range, Equation 9.9 has an integral to correctly sum the contribution of each probable value of the manufacturing line variables $\underline{\xi}$. As suggested in [Grossmann 91] this integral can be replaced by a summation by approximating the continuous jpdf with discrete probabilities leading to:

$$\begin{aligned} & min_{\underline{x}} \quad \sum_{i=1}^{j} w_i \cdot \left\{ \sum_{r=1}^{q} f(\underline{x}, \underline{\xi})\, p(\underline{\xi}) \right\} \\ & \text{subject to} \\ & \quad g(\underline{x}, \underline{\xi}) \le 0 \quad \forall \underline{\xi} \in \Xi \\ & \quad \underline{x} \in X_P \end{aligned} \tag{9.10}$$

On the other hand, for operating point variations, the range has no probability distribution. All points in the range are, in effect, equally probable and we need to insure correct operating over the entire range. In this case we need to find the minimum worst case point. This case leads to the following representation

$$\begin{aligned} & min_{\underline{x}} \quad \sum_{i=1}^{j} w_i \cdot \{ max_{\underline{\xi}} \quad f(\underline{x}, \underline{\xi}) \} \\ & \text{subject to} \\ & \quad g(\underline{x}, \underline{\xi}) \le 0 \quad \forall \underline{\xi} \in \Xi \\ & \quad \underline{x} \in X_P \end{aligned} \tag{9.11}$$

If we ignore the constraints in Equation 9.11, our formulation reduces to the *fixed tolerance problem* (FTP) [Madsen 79]. This formulation of the design problem was introduced for discrete device optimization in a world where discrete devices exhibited known tolerances (and unknown statistical variations). The lack of performance constraints in these approaches makes it difficult to apply these approaches to modern IC design problems. Therefore, the non-linear infinite programming formulation is more complete than the tolerancing problem formulation. In addition, the FTP formulation involved tolerances on the design variables ($\underline{x}$), while in our case the tolerances are on environmental and manufacturing line parameters that are orthogonal to design variables.

Our approach is to treat the manufacturing and operating point variations in the same manner. Therefore we apply the formulation in Equation 9.11 for both these kinds of variations. This can lead to pessimistic designs with respect to the statistical manufacturing variations since we employ the worst case point rather than weight each point with its appropriate probability. However this occurs only when the objective functions $f_i(\underline{x}, \underline{\xi})$ are dependent on $\underline{\xi}$. When the objectives are independent of $\underline{\xi}$, as in the case of active area of the design, the most common of the design objectives, there is no pessimism in design decisions.

Note that the *for all* constraints do not lead to pessimistic designs even though there is a jpdf in a fabrication line. This is because we model the yield as a constraint. If the 3σ contour is chosen, there is a constraint that $Y \geq 95\%$. In traditional yield optimization approaches, a 95% yield requires that the design acceptability region completely enclose the 3σ hyperellipse, just like in our formulation.

To overcome the pessimism resulting from the use of the minimax objective instead of the integral for the manufacturing line variations, we can use a post-design yield optimization approach to maximize the yield. In the future, the two types of variations can be treated separately to remove the pessimism arising from our common approach to these variations.

9.6 Summary

In this chapter we presented our complete variation tolerant formulation, which models the circuit design problem as a non-linear infinite programming problem. We began by describing the goals of our formulation, then introduced the range variable, and stated the formulation. The key idea of the range variable is that it naturally models variability in circuit operating points (*e.g.*, temperature, supply voltage, *etc.*) and can model a simplified form of statistical manufacturing line fluctuations (*e.g.*, MOS geometry, device threshold voltages) in a manner similar to earlier work on design tolerances. This simplification of a statistical jpdf for the manufacturing variations into range variables also incorporates a somewhat pessimistic form of yield maximization. We ended the chapter using a detailed example to show how this non-linear infinite program can model the complete design problem. We also used the example to highlight the important issues in this formulation. In the next chapter we will discuss methods for solving this formulation.

CHAPTER 10

Solving the Infinite Program

In this chapter we discuss the solution of the *non-linear infinite program* we formulated in the previous chapter. We will start by reviewing the infinite programming literature and then consider the specific approach we take in this book. A detailed description of the annealing-based implementation and a synthetic example using this implementation concludes the chapter.

10.1 Literature Review

In addition to our application, there are several other important applications for infinite or semi-infinite programming including non-linear fitting to measured data [Fiacco 83], earthquake resistant design [Bhatti 79], and control system design [Zakian 73]. As a result, although it is not as broadly applicable as linear, integer or non-linear programming, researchers have investigated various algorithms for solving infinite programs. In this section we will review these approaches.

We begin with a simplified form of the non-linear infinite program, with only a single design variable x, and a single range variable ξ. Also, we consider only a single objective, and a single constraint. We make these simplifications to simplify our explanation of semi-infinite and infinite programs, but the complete form will be discussed later in this chapter. With these simplifications, our problem is:

$$\begin{aligned} &\min_{x \in X_P} && f(x) \\ &\text{subject to} && g(x, \xi) \le 0 \quad \forall \xi \in \Xi \end{aligned} \tag{10.1}$$

Semi-infinite programming is the name used in the literature to refer to problems of the form shown in Equation 10.1. In this problem, a point x is feasible if $g(x, \xi)$ (as a function of ξ) is less or equal to zero on the *whole* of Ξ. If Ξ is a finite set, then $g(x, \xi) \le 0, \xi \in \Xi$ represents a finite number of inequalities, and the problem reduces to a common non-linear programming problem. On the other hand, if Ξ is an infinite set (as in the case of a continuous variable ξ), this problem is considered to be a *semi-infinite* program.

If, in addition to Ξ being an infinite set, the objective function is a function of ξ, then we have an *infinite program*, as shown in Equation 10.2.

$$\begin{aligned} &\min_{x \in X_P} && f(x, \xi) \quad \forall \xi \in \Xi \\ &\text{subject to} && g(x, \xi) \le 0 \quad \forall \xi \in \Xi \end{aligned} \tag{10.2}$$

The basic difficulty in solving this problem is the fact that the number of objectives is infinite, and the number of constraints too is infinite. A critical observation can be made to eliminate the *for all* term. We can replace the infinite number of objectives with a single objective function as determined by the problem $\max_{\xi \in \Xi} f(x, \xi)$. This is because the worst case value of the objective in x-space is the maximum value of $f(x, \xi)$ with respect to ξ. And, if we minimize the worst case objective, we are minimizing the objective for all ξ.

Similarly, we can replace the infinite number of constraints with a single constraint function as determined by the problem $\max_{\xi \in \Xi} g(x, \xi)$. Again, if we meet the constraint at the worst case value in ξ-space, we are meeting the constraint for all values in ξ-space. Note that the worst case value for $g(x, \xi)$, is the maximum value in ξ-space due to the direction of the inequality. If it had been a greater-than inequality, the worst case value would have been a minimum in ξ-space.

Thus the problem in Equation 10.2 is equivalent to

$$\begin{aligned} &\min_{x \in X_P} && \max_{\xi \in \Xi} f(x, \xi) \\ &\text{subject to} && \max_{\xi \in \Xi} g(x, \xi) \le 0 \end{aligned} \tag{10.3}$$

Therefore the objective and constraint are both no longer differentiable. This gives us an obvious first approach to solving the infinite programming problem: solve Equation 10.3 by applying methods of non-differentiable optimization [Polak 87].

A second approach to solving the original problem (Equation 10.2) is to overcome the infinite nature of the set Ξ by discretizing it to create a finite, approximate set. There are two basic approaches for this: *grid-based* and *cutting plane-based.* The simplest approach is to create a fine grid, and therefore a large but finite number of objectives and constraints, and solve for the optimum using standard non-linear programming approaches. A more elegant method is to create a coarse grid and solve a non-linear program with a few constraints and objectives. Then, we can use the information gained from this solution to further refine the grid in the region of ξ-space where this is necessary [Parsa 83].

Our solution approach is a derivative of the cutting plane-based discretization originally proposed by [Eaves 71]. We discuss this approach in detail in the next section.

10.1.1 A Conceptual Algorithm

In this section, we present a conceptual version of the cutting plane-based discretization algorithm on which our implementation is based. The starting point is, for simplicity, a semi-infinite program, as shown in Equation 10.4. Again we assume there is only a single (finite) objective, and a single constraint with an infinite set. We call this *Problem P*, and will refer back to this as we develop the conceptual algorithm.

$$P: \quad \begin{array}{ll} \min_{\underline{x}} & f(\underline{x}) \\ \text{s.t.} & g(\underline{x}, \underline{\xi}) \le 0 \quad \forall \underline{\xi} \in \Xi \\ & \underline{x} \in X_P \end{array} \tag{10.4}$$

The major difficulty in this problem is this infinite set Ξ. Again, we can introduce the worst case concept to replace this infinite set. Since we are trying to ensure that $g(\underline{x}, \underline{\xi}) \le 0 \ \ \forall \underline{\xi} \in \Xi$, the worst case point is the maximum value of g with respect to $\underline{\xi}$, as shown in Figure 10.1. If we ensure that this worst case point is less than zero, then all points in $\underline{\xi}$-space will be less than zero, and the constraint is satisfied for the infinite set.

Now, let $I(\underline{x})$ denote the problem shown in Equation 10.5:

$$I(\underline{x}): \qquad max_{\underline{\xi} \in \Xi}\, g(\underline{x}, \underline{\xi}) \tag{10.5}$$

A simple conceptual algorithm begins with a $\underline{\xi}_0 \in \Xi$ and solves Equation 10.6 to obtain $\underline{x}_0$.

$$P_0: \qquad \begin{array}{ll} min_{\underline{x}} & f(\underline{x}) \\ \text{s.t.} & g(\underline{x}, \underline{\xi}_i) \le 0 \quad i = 0 \\ & \underline{x} \in X_P \end{array} \tag{10.6}$$

(Note that P_0 has only one constraint, and so can be solved using standard constrained optimization approaches.) If we consider $\underline{\xi}_0$ to be the nominal point for the range variables, then P_0 corresponds to the problem for nominal synthesis.

Now, let $\underline{\xi}_1$ be the solution of $I(\underline{x}_0)$, *i.e.*, the solution of Equation 10.5 at $\underline{x}_0$. This gives Equation 10.7.

$$\begin{aligned} \underline{\xi}_1 &= I(\underline{x}_0) \\ &= max_{\underline{\xi} \in \Xi}\, g(\underline{x}_0, \underline{\xi}) \end{aligned} \tag{10.7}$$

Note that if $\underline{\xi}_1 = \underline{\xi}_0$ we have solved problem P and are done. In this case we already considered the worst case point in Ξ when solving P_0. Similarly, regardless of the value of $\underline{\xi}_1$, if the inequality constraint $g(\underline{x}_0, \underline{\xi}_1) \le 0$ is met, we have met the con-

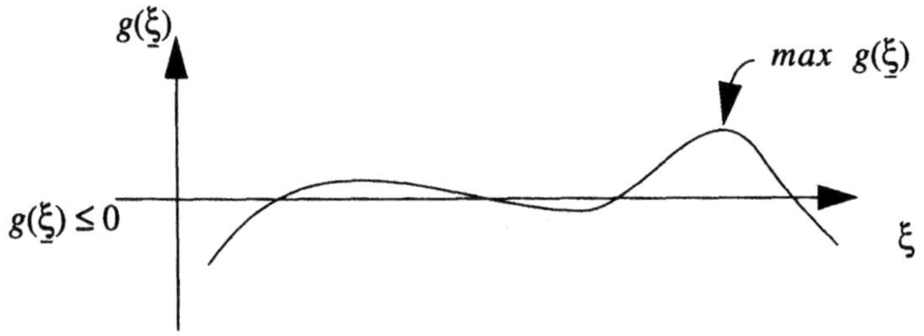

FIGURE 10.1 The Worst Case Point

straint, and are at the minimum value of the objective, implying we have solved the optimization problem.

However, if $g(\underline{x}_0, \underline{\xi}_1) > 0$, then we have not met the constraint, and we must then solve the problem in Equation 10.8, generating $\underline{x}_1$.

$$P_1: \quad \begin{array}{ll} \min_{\underline{x}} & f(\underline{x}) \\ \text{s.t.} & g(\underline{x}, \underline{\xi}_i) \le 0 \quad i = 0, 1 \\ & \underline{x} \in X_P \end{array} \tag{10.8}$$

The reason we consider both of the worst case points $\underline{\xi}_0$ and $\underline{\xi}_1$ in P_1 is to ensure that we will eventually converge. The subscript in the problem label indicates the iteration number. We can generalize this sequence of problems, labelled as P_ρ, in which there are only $\rho + 1$ constraints; they can be written as:

$$P_\rho: \quad \begin{array}{ll} \min_{\underline{x}} & f(\underline{x}) \\ \text{s.t.} & g(\underline{x}, \underline{\xi}_i) \le 0 \quad i = 0, 1, \ldots, \rho \\ & \underline{x} \in X_P \end{array} \tag{10.9}$$

This sequence is interspersed with the problems $I(\underline{x}_i)$ with $i = 0, 1, \ldots, \rho$. The flow of this conceptual algorithm is therefore a sequence of two alternating optimization problems as shown in Figure 10.2: a nominal problem, followed by a search for the worst case point over a range variable. Note that P_ρ only has a finite number of constraints. Therefore we have been able to convert the semi-infinite programming problem into a sequence of finite optimization problems.

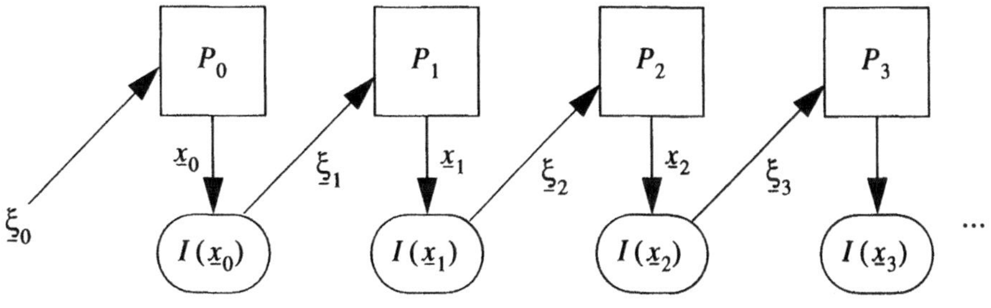

FIGURE 10.2 Flow of Conceptual Cutting Plane Algorithm

Let us consider the convergence properties of this approach. Consider the feasible set of the problems P_ρ:

$$F_\rho = \{\underline{x} | g(\underline{x}, \underline{\xi}_i) \le 0, i = 0, 1, \ldots, \rho\} \quad \textbf{(10.10)}$$

and Problem P:

$$F = \{\underline{x} | g(\underline{x}, \underline{\xi}_i) \le 0, \forall \underline{\xi} \in \Xi\} \quad \textbf{(10.11)}$$

The sequence of feasible sets satisfy Equation 10.12.

$$F_0 \supset F_1 \supset \ldots \supset F \quad \textbf{(10.12)}$$

Thus the sequence of objective function values have the property shown in Equation 10.13 (since more and more constraints are added as i increases):

$$f(\underline{x}_0) \le f(\underline{x}_1) \le \ldots \le f(\underline{x}_\rho) \le \ldots \le f(\underline{x}^*) \quad \textbf{(10.13)}$$

where $\underline{x}^*$ is any solution to Problem P. Therefore, this conceptual algorithm continues adding constraints (cutting planes for the linear problem) until the feasibility region reduces to the feasible set for Problem P, at which point the algorithm converges to determine $\underline{x}^*$.

One of the problems of using this approach, like any other discretization approach, is that as ρ increases, the number of constraints in Problem P_ρ increases, therefore we need to consider removing or pruning constraints for efficiency. This is often referred to as *constraint dropping*. Eaves and Zangwill, in [Eaves 71] show that the constraint $g(\underline{x}, \underline{\xi}_i) \le 0$ can be dropped from Problem P_ρ if:

- $f(\underline{x}_{\rho-1})$ is sufficiently larger than $f(\underline{x}_i)$, and
- $g(\underline{x}_{\rho-1}, \underline{\xi}_i) \le 0$ and
- the next solution $\underline{x}_\rho$ satisfies $f(\underline{x}_\rho) \ge f(\underline{x}_{\rho-1})$.

This constraint dropping scheme suggests that the growth of complexity of Problem P_ρ can be slowed if the objective increases monotonically. In the absence of convexity, however, the solution of P_ρ is only a local minima, so the sequence $\{f(\underline{u}_\rho)\}$ is not necessarily monotonically increasing, so the Eaves and Zang will constraint-dropping scheme cannot be applied, so we will need to use a different technique.

This study of the conceptual algorithm has shown that a solution to the non-linear infinite program is possible using a sequence of alternating sub-problems. These sub-problems are finite and can be solved by traditional optimization techniques. We have also seen that in the case of monotonically increasing $\{f(\underline{u}_p)\}$, the algorithm will converge, and constraint dropping may be used to improve the computational complexity of the algorithm. We will see how this is done in the domain of circuit synthesis in the next section.

10.2 Our Solution Approach

Our strategy is to formulate the manufacturable analog synthesis problem as a non-linear infinite programming problem and solve this formulation by extending our ASTRX/OBLX nominal synthesis system.

In the domain of circuit synthesis, complicated combinations of nonlinear device characteristics make it extremely difficult, if not impossible, to make any statements about convexity. Hence, as with our nominal synthesis formulation, we use a simulated annealing optimization method, and exploit its hill-climbing abilities in the solution of the non-linear infinite programming problem.

In this section, we describe our approach for incorporating infinite programming into ASTRX/OBLX. We first present the basic approach we use in solving the infinite programming problem, and then discuss the issues generated by this approach in more detail.

10.2.1 Basic Approach

The basic approach we use is the sequence of optimization sub-problems highlighted in the conceptual example seen in Section 10.1.1. To understand this further let us consider a simple example in the circuit synthesis domain.

In the example we have chosen, the optimization problem can be written as in Equation 10.14, with one (infinite) objective function which is $power(x, V_{dd})\ \ \forall V_{dd} \in [4.5V, 5.5V]$, one (again infinite) constraint which is $power(x, V_{dd}) - Pspec \le 0\ \ \forall V_{dd} \in [4.5V, 5.5V]$, a single design variable, and a single range variable. Again for simplicity we assume no state variables, no equality constraints, and just a single design variable. So, our problem is formally:

$$\begin{array}{lll} min_x & power(x, V_{dd}) & \forall V_{dd} \in [4.5, 5.5] \\ \text{s.t.} & power(x, V_{dd}) - Pspec \le 0 & \forall V_{dd} \in [4.5, 5.5] \\ & x \in [X_{min}, X_{max}] & \end{array} \quad \textbf{(10.14)}$$

In our conceptual algorithm, we started off with $\xi_0 \in \Xi$, which in the domain of circuit synthesis maps to the nominal operating point and device parameters for the manufacturing process. Therefore for this example, we start with the nominal point, *i.e.*, $V_{dd} = 5.0V$. In the conceptual algorithm we next solved the problem P_0. This is the nominal synthesis problem, since it ignores all the other values of V_{dd}. Graphically, we can depict this problem as in Figure 10.3, with the optimal solution x_0.

However, the real problem is more complicated, as shown by the cost function in Figure 10.4. In addition to the original design space in the x dimension, there is an additional space in the V_{dd} dimension arising from the *for all* constraint. In our conceptual algorithm we presented a sequence of alternating optimization sub-problems

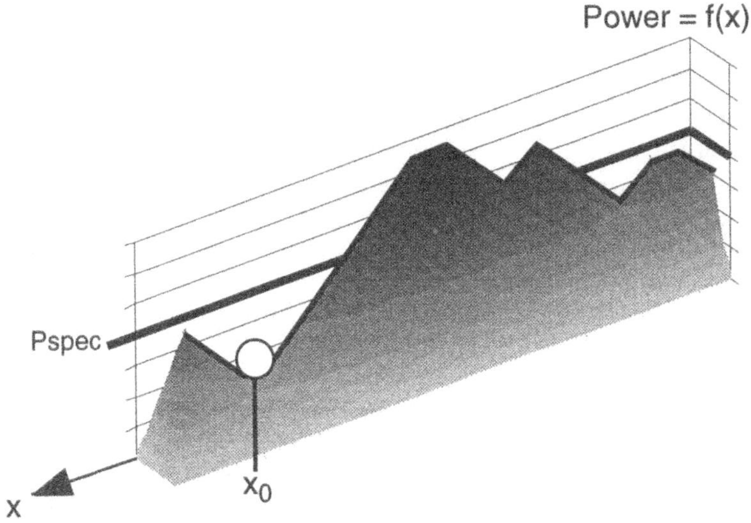

FIGURE 10.3 The Nominal Cost Function, $V_{dd} = 5.0V$

to help overcome this impediment. In the domain of circuit synthesis, this approach involves using the nominal approach in ASTRX/OBLX in its entirety to do the circuit synthesis. However, as the circuit within ASTRX/OBLX evolves, we periodically freeze the circuit design variables x, thus stopping ASTRX/OBLX, and go solve a new optimization problem to determine the worst case value of the range variable. Thus we have two optimization problems, P_ρ (Equation 10.9) solved by ASTRX/OBLX, and a new optimization problem $I(\underline{x})$ (Equation 10.5).

10.2.2 Outer Optimization

The problem P_ρ is called an *Outer Optimization* problem. In this optimization the worst case variables are held constant, and the decision variables in the optimization are altered with the aim of finding the best design for the specifications at the set of *a priori* known worst case values. By *a priori* we mean before the ρth outer optimization begins.

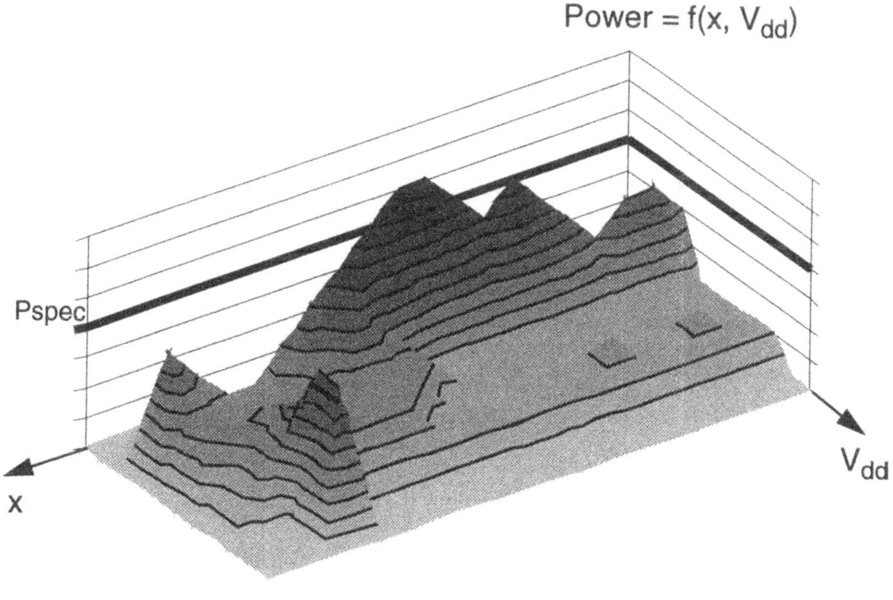

FIGURE 10.4 The Meaning of $\forall V_{dd} \in [4.5, 5.5]$

At this point in our example, we have just solved the first outer optimization problem, the nominal synthesis problem P_0, with ASTRX/OBLX. We will be solving subsequent outer optimization problems later. For now, let us continue to look at the first inner optimization problem.

10.2.3 Inner Optimization

The problem $I(x)$ is called an *Inner Optimization* problem. In this problem, the design decision variables are held constant, and an optimization is done in the range variable space for the worst case value of the range variables.

This was the second step in the conceptual algorithm: solve $I(x_0)$. The cost function of this first inner optimization sub-problem is shown in Figure 10.5. Suppose the outer optimization finds the minimum x_0. We freeze the design variables at $x = x_0$ and take the slice of $power(x = x_0, V_{dd})$ (shown in Figure 10.5) as the cost function for the inner optimization. Since we want to ensure that $power \leq Pspec$, this corresponds to a maximization problem. If the maximum value of power in the

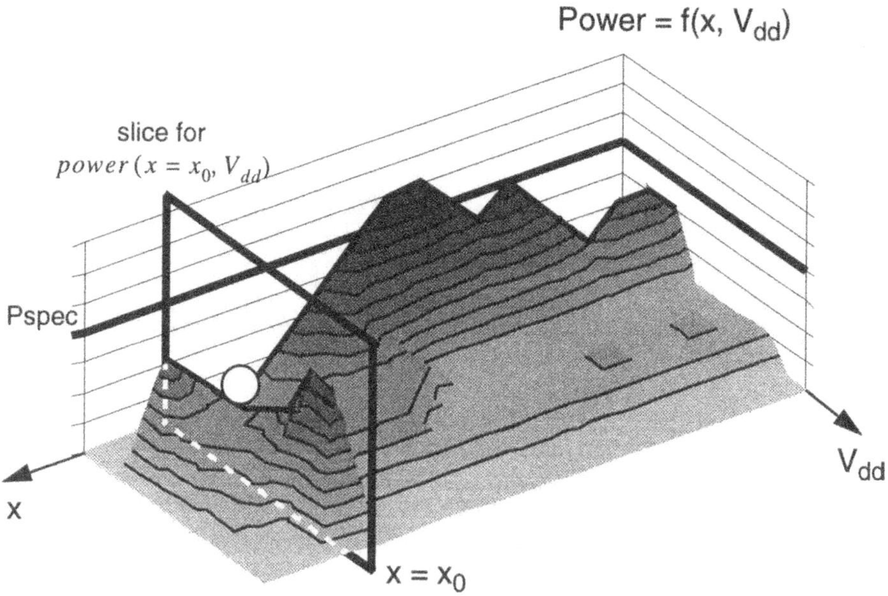

FIGURE 10.5 Inner Optimization Cost Function

$power(x = x_0, V_{dd})$ cost function meets the specification, then all points across the V_{dd} axis will meet the specification. Thus the responsibility of the inner optimization is to find the worst case point, V^*_{dd} as in Figure 10.6.

In both the conceptual description and in this example there is only a single optimization sub-problem in each inner optimization in the sequence of inner optimizations. This is due to the single constraint semi-infinite program used to describe the conceptual algorithm. In the case of the power minimization example, this is because we have the same function in the objective and the constraint. If there are multiple infinite constraints in the original problem (*i.e.*, $\underline{g}$ is a vector of functions, not a single function), each of the inner optimizations $I(x_i)$ will involve more than one optimization sub-problem. There will be one sub-problem for each objective or constraint, since we need to determine the worst case point for each objective or constraint.

Now that we have the worst case point V^*_{dd}, we need to link this back to the second outer optimization.

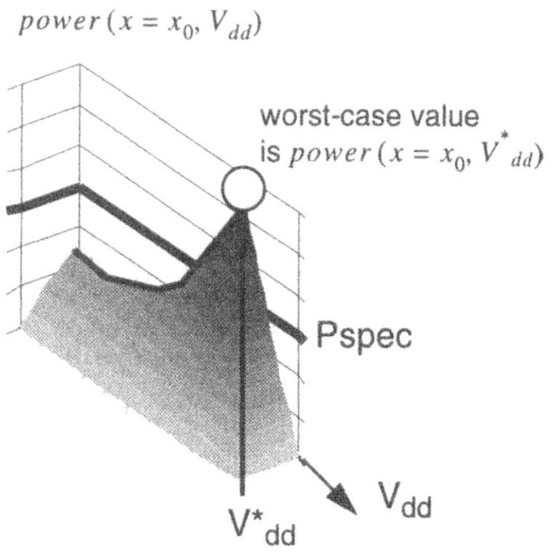

FIGURE 10.6 Solution of the Inner Optimization (Slice from Figure 10.5)

10.2.4 Sequence of Inner and Outer Optimizations

In the conceptual algorithm of Section 10.1.1, the link back to the second outer optimization involves adding another constraint to the finite problem P_0 to create another finite optimization problem P_1. Since the nominal ASTRX/OBLX solves the constrained optimization problem by converting it to an unconstrained optimization problem, the addition of V_{dd}^* is handled by altering the ASTRX/OBLX cost function.

The effect of adding this information must modify the cost function of the outer optimization in such a way to prevent it from searching around the areas that are optimal in the *nominal* sense, but not in the *for all* sense implied by the range variables. Figure 10.7 shows this effect of adding the result of the inner optimization to the outer optimization cost function for our simple example about circuit power. The effect of this *critical* point or *worst-case* point (V_{dd}^* from Figure 10.6) is to make the original outer optimization result (x_0) sub-optimal. Thus we need to run the outer optimization again on this new cost function. If the inner optimization does not add any new critical points, then the outer optimization cost function remains the same, hence no further outer optimizations are necessary.

We generate the second outer optimization cost function shown on the right graph of Figure 10.7 using the worst-case points collected so far: V_{dd}^{nominal} and V_{dd}^*. This cost function can be written as in Equation 10.15:

$$
\begin{aligned}
C(x) &= max\{power(x, V_{dd}^{\text{nominal}}), power(x, V_{dd}^*)\} + \\
&\quad max\{0, power(x, V_{dd}^*)\} + \\
&\quad max\{0, power(x, V_{dd}^{\text{nominal}})\}
\end{aligned}
\tag{10.15}
$$

The first term in the outer optimization cost function relates to the discretization of the infinite objective. In the nominal case, we were trying to minimize $power(x)$, and now we are minimizing the worst case value of the power at the discretized points. The second and third terms relate to the discretization of the infinite constraint. In the nominal cost function we have only one term for the constraint; now we have a term for every discretized value of V_{dd}. Solving this cost function, in terms of ASTRX/OBLX, is like designing a single circuit that is simultaneously working at two different operating points, V_{dd}^{nominal} and V_{dd}^*.

10.3 Local and Global Optimization Methods

Selecting the best optimization engine to solve the two optimization problems is not an easy task. Both local and global optimization methods have been used for the outer optimization problem.

A local minimum of a cost function $f(x)$ is a point x^* at which the function exhibits zero slope ($f'(x) = 0$) and non-negative curvature ($f''(x) \geq 0$). The earliest numerical optimization approaches used gradient-based optimization methods that determine a local minimum in the cost function (and, hence, are starting point dependent, like DELIGHT.SPICE [Nye 88]).

However, since we are extending our nominal formulation, the most straightforward approach is to continue to use simulated annealing as our optimization algorithm. In fact we choose simulated annealing for both optimizations. First, since it is tightly integrated with ASTRX/OBLX, the starting point of our strategy, we use annealing to solve our outer optimization problem. Second, since the inner optimization cost function is computed in the same manner as the outer optimization cost function, we also choose annealing to solve our inner optimization problem. We refer to this as an *annealing within annealing* formulation. The first annealing refers to the inner optimization annealing, and the second annealing refers to the outer optimization annealing.

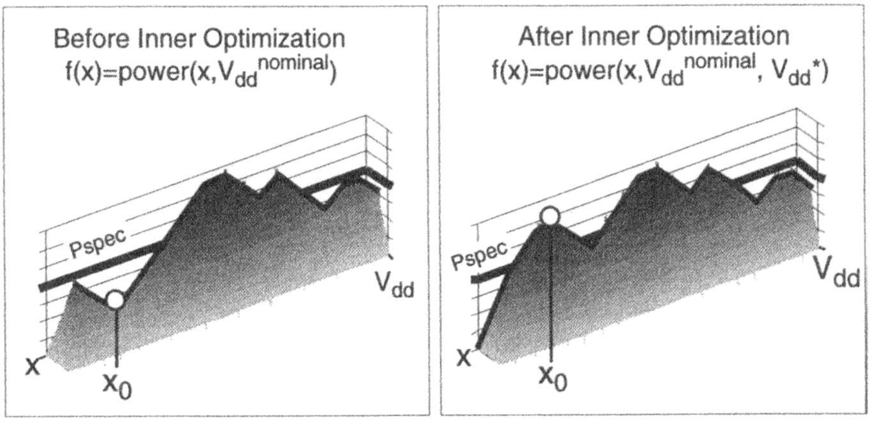

FIGURE 10.7 Effect of Inner Optimization on Outer Optimization Cost Function

10.4 Annealing in Annealing Implementation

So far we have been considering an approach in which a complete outer optimization is done, then a complete inner optimization, then back to a complete outer optimization and so forth until we converge to a solution. Recall that a complete outer optimization, for us, involves a complete nominal synthesis run to design a circuit that simultaneously meets the performance specifications at all the worst case points considered for that outer optimization.

However, a nominal synthesis system like ours cannot be re-rerun with a new constraint for each critical point identified by the inner optimization. This is mainly because of the long run times required for the nominal design. As we discussed in Chapter 6, the reported run times range from minutes to hours depending on the size of the circuit being synthesized. An alternate, more fine-grained, approach for the interaction between the outer and inner optimization is needed in this framework. This fine-grained interaction, we call the *inner annealing* within the *outer annealing*, resulting in an annealing within annealing implementation.

Recall that we are trying to determine how to interface the inner optimization with our nominal synthesis method in order to solve the non-linear infinite programming model of circuit synthesis. There are three possible approaches based on the description of the annealing algorithm we gave in Section 4.1:

1. Let the entire nominal synthesis be the outer optimization. We therefore run an entire nominal synthesis circuit synthesis to completion, then use the inner optimization to determine critical points that can be used to update the performance constraints in nominal synthesis, and then rerun the nominal synthesis again. This quickly becomes impractical due to the length of each nominal synthesis run.
2. Let the per move iteration within nominal synthesis be the outer optimization. Therefore after each move (Line 4 in Figure 4.1 on page 82), we execute an inner optimization which generates a new critical point. We then modify the nominal synthesis cost function appropriately, and return back to the per move iteration within nominal synthesis. Such a fine grained interaction between the inner and outer optimizations is also not reasonable. Note that each inner optimization involves several iterations, or moves of the $\underline{\xi}$ range variables; thus, we would have one nominal move generation followed by several moves of the range variables, which is clearly unbalanced. The circuit design variable space is the larger and more complicated space, hence more effort should be expended in determining the optimum in that space rather than determining more and more critical points.

3. Let the per temperature iteration within nominal synthesis be the outer optimization. In this approach, the circuit design variables $\underline{x}$ are modified several times (as controlled by the annealing parameters) between each inner optimization. In each inner optimization, the range variables $\underline{\xi}$ are modified several times until the worst-case or critical point is determined. This provides an ideal interaction point between the outer and inner optimizations.

In practice, choosing the per temperature iteration means that on the order of hundreds of inner optimizations take place during a single synthesis, compared to hundreds of thousands if the per move iteration was chosen. Therefore, the total run time for essentially similar results is least when the per temperature iteration is chosen.

10.4.1 Monotonicity in Inner Optimization

We motivated the need for an optimization process to determine the worst-case points using Figure 8.4 on page 194, where the worst-case point for the operating point range occurs within the bounds of the range rather than at the corner. Sometimes, however, the worst case point does occur at a corner of the range used to model the operating point range or the sufficiently probable range of the manufacturing variations, and optimization is not necessary to determine this worst case point.

Therefore, for computational efficiency, the first step of the inner optimization involves checking to determine if full nonlinear optimization is really necessary. This is done by creating a model of the $C(\underline{x}_i, \underline{\xi})$ cost function, where $\underline{x}_i$ is the configuration currently being considered by the outer optimization. We use large-signal derivatives of the cost function with respect to $\underline{\xi}$ to create this model, and then estimate empirically if the model is monotonic or not. If the model is monotonic, the critical point will occur at a corner, and no optimization is necessary. If the model is not monotonic we use an annealing-based optimization process to determine the worst case point.

By "large-signal derivatives" we mean evaluating the derivative of the cost function $C(\underline{x}_i = \mathit{fixed}, \underline{\xi})$ with respect to $\underline{\xi}$. We determine this by evaluating the cost at the nominal and end-point values for each of the worst-case directions in the vector $\underline{\xi}$, and for each direction comparing these three numbers (low end cost, nominal cost, and high end cost) to estimate monotonicity.

10.4.2 Constraint Pruning

Constraint pruning or constraint dropping can be employed to reduce the total run times of the combined outer/inner optimization. Recall that each inner optimization has the effect of indicating to the outer optimization that it should simultaneously design the circuit for another value of the range variables. For the outer optimization, this means that each configuration move (change in the design variables, $\underset{\sim}{x}$) needs one more complete circuit simulation. This is the primary source of growth in terms of run-time for this approach.

To control this growth in run-time, a constraint pruning method can be applied. In our non-convex domain, the Eaves and Zangwill [Eaves 71] approach discussed in Section 10.1.1 cannot work. Instead, we consider dropping constraints using a *least recently inactive* heuristic.

During the early stages of the annealing, the Metropolis criterion encourages search across the entire design space, leading to the addition of several constraints that are not useful during the later stages of more focussed searching. Our approach is to collect statistics for each constraint to indicate how often it is active. A constraint is considered to be active if the constraint is violated. In the example of power minimization, we can look at Equation 10.15 on Page 230. Each time a constraint is active, the *max()* function will return a positive number to increase the cost at that point. If the constraint is inactive, the *max()* function returns a zero, leaving the cost unaffected by that term. If a constraint is inactive for a long time, where "long" means that many circuit configurations have been generated and evaluated in the outer optimization, one of two situations can occur. Either another constraint has been added that is more strict than the inactive constraint, or the search in x-space has evolved to a region in this space where this constraint is no longer active.

In either case, we can drop this inactive constraint without affecting the cost function. There is a danger to doing this: the constraint may become active at a later point. We therefore need to regularly check for this, which we do during each per temperature iteration of the outer annealing. Therefore our approach is not to drop a constraint *completely*, but rather to *suppress* the evaluation of the constraint. This saves us the time-consuming component of the worst-case point, the performance evaluation needed in each outer move. In addition, late in the annealing, we do not suppress any evaluation, and all worst-case points are evaluated. This is to ensure that the final answer will meet all the constraints generated by the inner annealing, and hence will be a variation tolerant analog cell.

10.5 Synthetic Example

Our first test after implementing the above approach for solving infinite programs is to consider a suite of infinite programs suggested in [Coope 85]. In this section, we will illustrate the algorithm by applying this implementation to one of these problems. Since this is a synthetic problem, it shows the important characteristics of the coupled outer/inner optimization approach. We will present actual results on circuits in the next chapter. This synthetic problem is a small convex continuous semi-infinite programming problem. Again we use a semi-infinite program as a simplified version of an infinite program to illustrate how the annealing within annealing scheme works.

The problems in the benchmark suite of [Coope 85] are all of the form shown in Equation 10.16.

$$\begin{array}{ll} \min_{\underline{x}} & z = f(\underline{x}) \\ \text{s.t.} & \begin{cases} g(\underline{x}, \underline{\xi}) \le 0 \\ \forall \underline{\xi} \in \Xi \\ \underline{x} \in \Re^{x} \end{cases} \end{array} \tag{10.16}$$

The problem we will solve is reproduced below:

$$\begin{aligned} f(x_1, x_2) &= \frac{1}{3}x_1^2 + x_2^2 + \frac{1}{2}x_1 \\ g(x_1, x_2, \xi) &= (1 - x_1^2\xi^2)^2 - x_1\xi^2 - x_2^2 + x_2 \\ \Xi &= [0,1] \end{aligned} \tag{10.17}$$

A graph of the elliptical contours of $f(x_1, x_2)$, the contours of the constraint as described by the set $\{g(x_1, x_2, \xi_i) = 0,\ \xi_i = \{0, 0.1, \ldots, 0.9, 1\}\}$ and the final feasibility region for the problem is shown in Figure 10.8 to give a visual description of the problem (white means feasible).

Figure 10.9 shows the evolution of the objective (f) and constraint (g) terms in the annealing cost function as a function of the cooling iterations (*i.e.*, time) for the solution. In the case of the semi-infinite programming formulation of the circuit design problem, there is at least one given point in the ξ-space (the nominal design point). Our implementation of the NLIP solver requires this initial constraint too. Since this particular problem does not have one, the middle of the range *i.e.*, $\xi = 0.5$, is used as the nominal point. This is indicated by the arrow from the left-most feasibility graph to the annealing dynamics graph, where the white region is feasible. In this particular run, we have tried to reduce the run-time by checking for monotonicity in the ξ-space.

The penalty term in the annealing cost function corresponding to the inequality constraint evaluates to zero if the design point for the current iteration $\underline{x}_i$ is in the feasible region, and to the actual constraint function value if infeasible. In the first iteration (at the end of the first temperature) the monotonicity check indicates that the constraint function does not vary with ξ (*i.e.*, we are in the feasible region). After two iterations, the annealing visits the states of $\underline{x}$-space around the neck of the feasibility region for $g(x_1, x_2, 0.5)$. The monotonicity check indicates that the upper bound to the ξ-space is the worst-case point. Therefore this is added to the list of constraints; this is shown in Figure 10.9 with the arrow from the middle feasibility graph to the annealing dynamics graph. From now on, both $g(x_1, x_2, 0.5)$ and $g(x_1, x_2, 1.0)$ are evaluated (*i.e.*, we have two constraints). After 6 cooling steps in the outer annealing algorithm, the third constraint $g(x_1, x_2, 0.0)$ is added (in Figure 10.9 shown by the arrow from the right feasibility region to the annealing dynamics graph). No other worst-case points occur in the rest of the run, which produces the correct answer for this problem.

Results for this problem (as well as other sample problems) appear in Table 10.1. Each problem occupies three rows in the table. The first row is the correct answer, reproduced here from [Coope 85], the second row contains results of an optimization

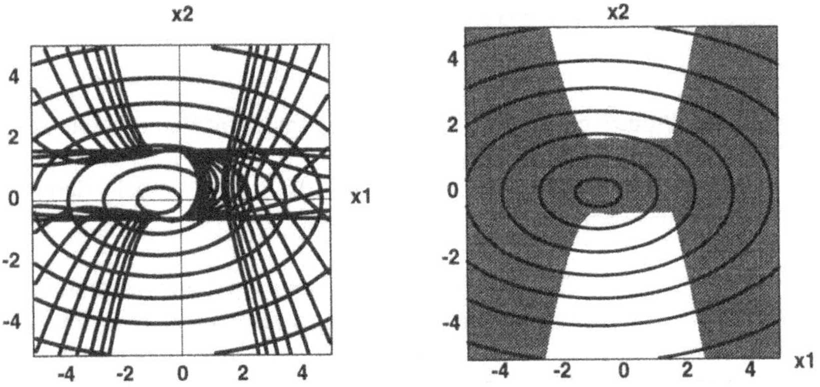

FIGURE 10.8 Feasibility Region for the Synthetic Problem

run in which annealing in the inner optimization was avoided by applying the test for monotonicity, and the third is results for the run in which the full inner annealing is done at each temperature. The optimum point generated by the run, the value of the objective at the optimum, the set of worst-case constraints added during the run, the CPU time in solving the problem and the number of annealing iterations (moves) are shown. In general, for such synthetic problems, we have seen that either approach can determine the correct set of worst-case constraints and get to within 1% of the exact optimum. The monotonicity test usually eliminates the time required for the inner annealing, and hence usually reduces the CPU time. However, this is not the case in the first example because of the stochastic nature of outer annealing (which in the monotonic case requires more moves than does the outer annealing in the annealing within annealing run).

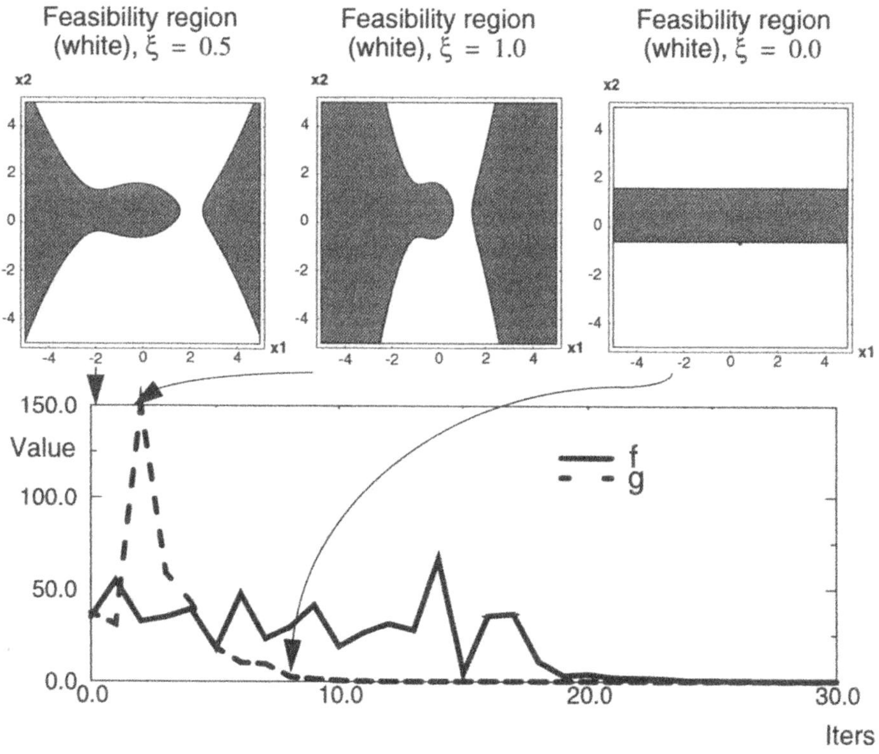

FIGURE 10.9 Sequence of Incorporation of Worst-Case Constraints into Outer Optimization

10.6 Summary

In this chapter we presented the details of the coupled outer/inner optimization approach we use to solve the non-linear infinite programming formulation of the circuit synthesis problem. We first reviewed ideas presented by researchers in infinite programming, and highlighted a conceptual algorithm to solve the infinite program. We then presented our coupled outer/inner optimization approach, which we explained using a simple graphical example. We then looked back at the annealing algorithm used in our nominal synthesis approach to determine how to couple the outer and inner optimizations. Finally we applied the implementation on a synthetic benchmark infinite programming problem as a pre-cursor to the actual circuit synthesis results we will present in the next chapter.

TABLE 10.1 Results of Annealing-based Non-Linear Infinite Programming Method

Soln Method	$\underline{x}^*$	f^*	ξ_{active}	CPU Time (s)	Number of Moves
1 exact	(-0.75, -0.618)	0.194	{0, 1}	-	-
1 mono	(-0.7536, -0.6183)	0.195	{0.5, 1, 0}	80	153010
1 anneal	(-0.7567, -0.6201)	0.197	{0.5, 0.9998, 0.0003}	68	113394
2 exact	(-1, 0, 0)	1	{ (0, 0) }	-	-
2 mono	(-1.000, 0.0452, 0.0673)	1.006	{ (0.5, 0.5), (1, 1), (0, 0) }	108	349662
2 anneal	(-0.989, -0.0369, -0.0615)	1.056	{ (0.5, 0.5), (0.84, 0.95), (0.008, 0.038) }	154	362546
3 exact	(0.719961, -1.450487)	97.16	{ 0 }	-	-
3 mono	(0.71198, -1.4506)	97.24	{ 0.5, 1, 0 }	40	122064
3 anneal	(0.71138, -1.4445)	97.26	{ 0.5, 0.995, 0.002 }	53	126034

CHAPTER 11

Variation-Tolerant Synthesis Results

In this chapter we present results to show that our non-linear infinite programming formulation can actually synthesize variation-tolerant analog cells. To test our variation-tolerant formulation, we created ASTRX/OBLX NLIP, named to show it is a non-linear infinite programming extension of the nominal version of ASTRX/OBLX. To avoid confusion, in this chapter we typically refer to the nominal version of the tool as simply ASTRX/OBLX, while calling the variation-tolerant version NLIP. To show the advantages of the NLIP formulation when compared to even the best of nominal synthesis results, we resynthesize several circuits from Chapter 6 with attention to operating point ranges and simplified manufacturing line variations. The circuits range from simple five-transistor OTA cells to forty-four transistor folded cascode op-amps.

We also present results for a new type of circuit that could not be synthesized prior to the variation-tolerant formulation. This new circuit is a band-gap reference circuit. Since this circuit requires temperature insensitivity, it is an example of a class of circuits that cannot be designed at all by nominal synthesis techniques (which lack any knowledge of sensitivity to variables like operating temperature since they are not in the list of design variables).

In our first two results, we consider only the V_{dd} operating range in addition to nominal synthesis. Next we consider a synthesis example in which simplified manufacturing line variations are taken into account. Finally, we consider the band-gap reference circuit.

11.1 Simple OTA Circuit

As our first circuit example, we consider the Simple Operational Transconductance Amplifier circuit originally shown in Figure 6.1 and repeated in Figure 11.1. There are 6 design variables: the width and length of the differential-pair transistors (M1 and M2), the width and length of the current mirror transistors (M3 and M4), the

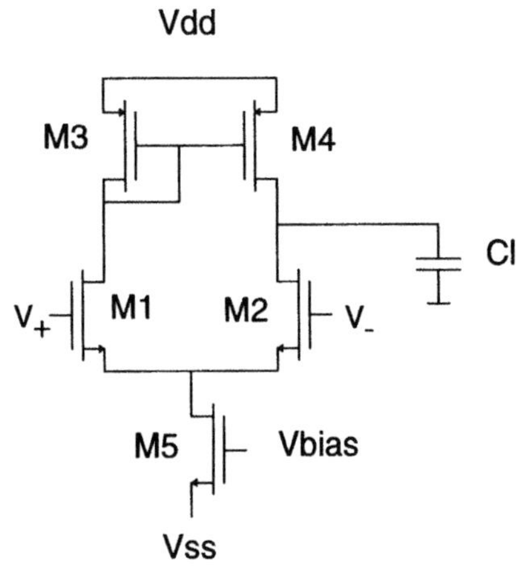

FIGURE 11.1 An Operational Transconductance Amplifier

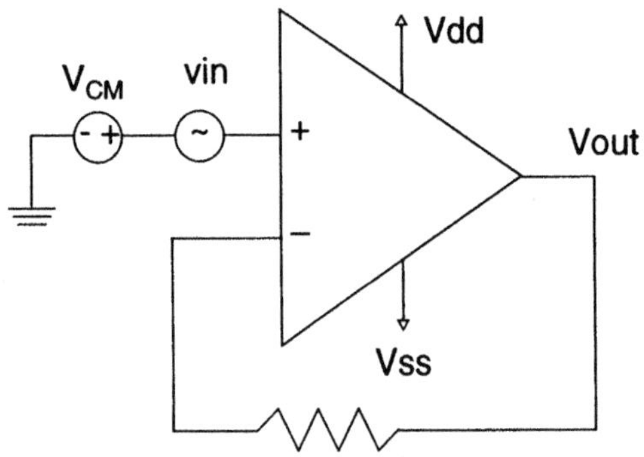

FIGURE 11.2 Test-jig to Measure D.C. gain of Operational Transconductance Amplifier

width and length of the tail current source transistor (M5), and the Vbias voltage. For the NLIP formulation, we added the V_{dd} operating point variable as a worst-case variable.

We compare the designs generated by looking at the performance graphs across the $4.5 \le V_{dd} \le 5.5$ operating range using the circuit of Figure 11.1 as the test-jig. Figure 11.3 shows the D.C. gain performance of the ASTRX/OBLX and NLIP formulation design, simulated at $V_{CM} = 2.5V$ (labeled Nominal) and $V_{CM} = V_{OH}$, the specified output high voltage (3.75V). For $V_{CM} = 2.5V$, note that at the nominal operating point ($V_{dd} = 5.0V$) the D.C. gain of the ASTRX/OBLX design is no more sensitive to the operating point than is the NLIP design. This illustrates that even adding small-change sensitivity constraints at the nominal operating point would not improve the design. The actual worst-case gain of this circuit will occur when the common mode of the input voltage (called V_{CM} here) is at its highest specified value, in this case the highest output voltage (V_{OH}) since the test-jig is configured for unity-gain feedback and V_{dd} is at its lowest value. It is clear from the graph that it is necessary to use the NLIP formulation to ensure that the D.C. gain is relatively insensitive to the operating range. Although the graph in Figure 11.3 suggests that the worst case point is in a single corner of the range, this is actually not the case in this example. Note that we have only presented one performance measurement, the D.C. gain. If one considers the other specifications requested for this circuit (linear specifications

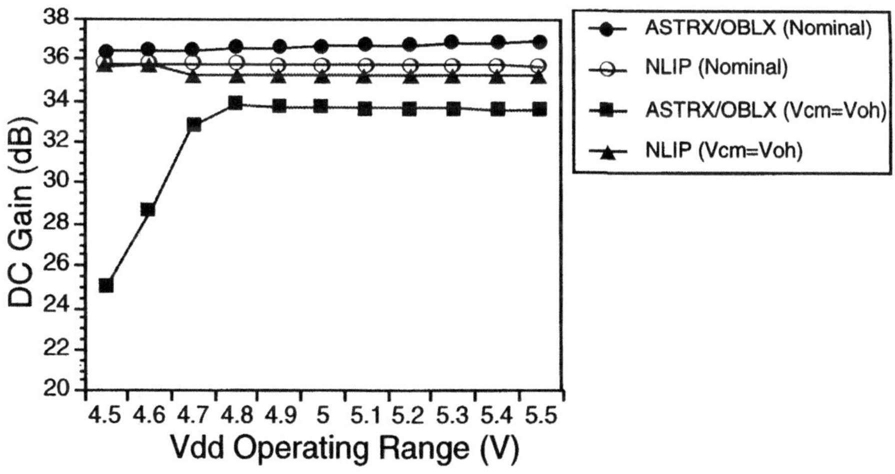

FIGURE 11.3 OTA D.C. gain Performance

like unity gain frequency, phase margin, power supply rejection ratio, common mode rejection ratio, and non-linear performances such as output high voltage and output low voltages, as well as non-linear transient performances such as slew rate), the worst case point occurs in the other corner. In one case, the worst case point occurs simultaneously at *both* the corners. In this simple example there were no worst case points internal to the range, at the final design point $\underline{x}^*$; however, since we visited a sequence of designs in our path to this final design point, we actually needed to consider the potential for interior worst case points during the synthesis process.

11.2 Folded Cascode Amplifier

We repeated the above experiment with a larger, more complicated circuit: the folded cascode amplifier. The original circuit schematic was shown in Figure 6.4, and is shown with slight modifications in Figure 11.4. In this design there are twelve designable widths and lengths, a designable compensation capacitance and two D.C. bias voltages. To this design, we added the V_{dd} operating range variable, just as in the previous experiment.

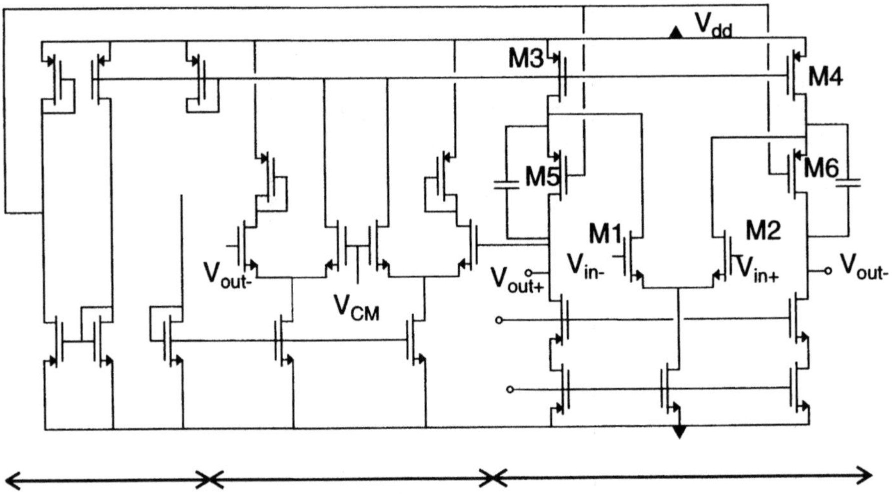

FIGURE 11.4 The Folded Cascode Amplifier

We simulated the nominal ASTRX/OBLX and the NLIP designs' output swing across the V_{dd} operating range using the test-jig shown in Figure 11.5, and the results obtained are shown in Figure 11.6. The output swing of the amplifier is a strong func-

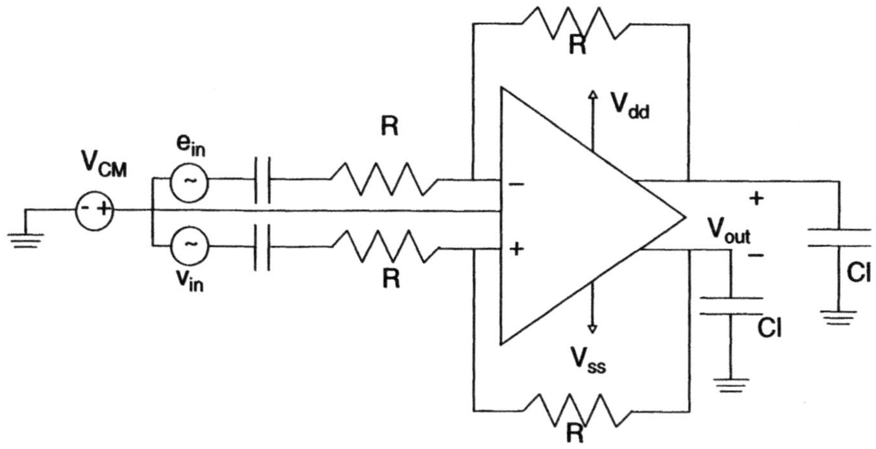

FIGURE 11.5 Test-jig for Folded Cascode Amplifier

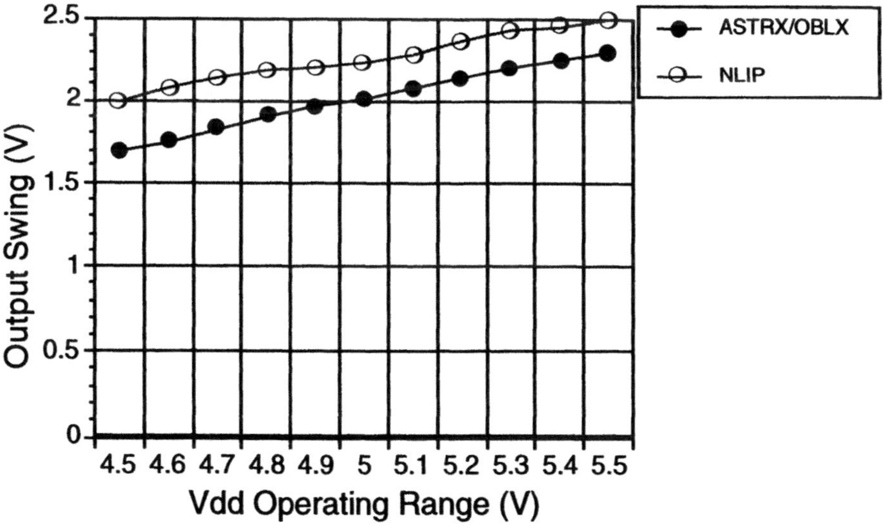

FIGURE 11.6 Folded Cascode Output Swing across V_{dd}

tion of the region of operation of the output transistors (M3, M4, M5 and M6). The output swing is obtained by using a large-signal A.C. input, and determining the output voltage at which the output transistors move out of saturation (which will cause clipping in the output waveform). Compared to the OTA, this is a much more difficult design, hence the output swing specification of 2.0V is just met by both the nominally synthesized design of Chapter 6 (at the nominal power supply voltage of 5.0V) and the NLIP design (across the entire operating range).

The nominal synthesis result fails to meet the output swing specification for the lower half of the operating range ($V_{dd} < 5.0V$). This is an inherent problem in a nominal synthesis formulation. For an optimal design, it biases the circuit so that the output transistors are at the edge of saturation, and a slight decrease in the V_{dd} voltage results in their moving out of saturation, hence the output swing falls below the 2.0V specification.

11.3 Manufacturing Line Variations

In our next experiment we again reconsider the OTA circuit of Figure 11.1. This time we introduce a simplified model of manufacturing line variations into the synthesis problem. One of the common problems faced during statistical design is the lack of appropriate statistical models of the fabrication line. To overcome this problem we pick a few variations which are considered most critical and assume that they are statistically independent. These variations are global variations in transistor geometries, global parametric variations in threshold voltages, and intra-chip parametric variations in the threshold voltages of matched devices. The practical effect of these assumptions is to transform the manufacturing problem into a tolerancing problem in the style of [Madsen 79].

In this run there are six circuit design variables, and six range variables to model the manufacturing line disturbances. The model we use for this run is the BSIM1 [Sheu 87] model which has about 80 device model parameters. [Krishna 95a] shows that only the flat-band voltage, V_{FB}, oxide thickness, t_{ox}, and the width and length reduction parameters are significant. Of these, t_{ox} is very tightly controlled in modern processes, hence is unlikely to be a source of parametric yield loss in small analog circuits. In addition, since device widths are much larger than the lengths in analog circuits, we can ignore the width reduction parameter's effect compared to the rest of the set. Hence we retain just the flat-band voltage, V_{FB}, and the length reduction parameter, ΔL. Since this circuit uses identical pairs of devices, we need to consider the intra-chip variation in addition to the inter-chip variation, ΔV_{FB}. We therefore choose six parameters, three each for the p and n devices (V_{FB}, ΔL, ΔV_{FB}). In addition, we use statistical independence between these variations and set ranges for each of the 6 variations using common-sense values for their expected standard deviation as shown in Table 11.1 . For the inter-chip variations, this was a constant number. For the intra-chip variations the standard deviation will be a function of the circuit layout, both in terms of device geometry as well as device separation. Instead of hard-coding a formula for this dependance, we choose instead to provide a flexible modeling interface in which the designer can use any function of device geometry and separation. For this experiment, we use a model suggested by [Pelgrom 89] which determines the standard deviation of the mismatch as a function of the device area.

The global device geometry variation used is a variation on the ΔL parameter used in determining the effective device length, $L_{eff} = L - \Delta L$, where L is the drawn length of the device. We use a range with a non-zero median to model this as an asymmetric variation. V_{FB}, the flat-band voltage of the p and n devices is the other global parametric variation in this experiment. We use a zero-mean variation which is added to the nominal V_{FB} generated by the BSIM device model equations in the encapsulated device evaluator. The ΔV_{FB} intra-chip variation term is used to model the expected mismatch in the flat-band voltages for the matched devices (between M1 and M2, and between M3 and M4). Note, the intra-chip parametric variations are particularly challenging because their amplitude depends on the design variables — these variations are roughly proportional to $1/(\sqrt{WL})$. The numbers given in the ranges shown in Table 11.1 are the proportionality constants, thus the actual range changes as the design is modified.

During actual synthesis, we expect to see that the device sizes will be larger to minimize the effect of the mismatch in geometry and threshold voltage. It should be obvious that larger geometries reduce the sensitivity to the ΔL variation. In addition,

TABLE 11.1 Variations used in Manufacturing Line Variation Experiment

Variation	range for p device	range for n device
ΔL	$[-0.5\mu m, -0.2\mu m]$	$[-0.4\mu m, -0.1\mu m]$
V_{FB}	$[-0.1V, 0.1V]$	$[-0.2V, 0.2V]$
ΔV_{FB}	$[-1 \times 10^{-10}, 1 \times 10^{-10}]$	$[-12 \times 10^{-9}, 12 \times 10^{-9}]$

larger devices imply smaller ranges for the ΔV_{FB} mismatch due to the [Pelgrom 89] model used, again pushing for larger designs.

Figure 11.9 shows the random component of the input referred offset voltage (the systematic component of both designs is less than 0.5mV), at the 64 worst-case corners of the 6 variables used in the design. This graph was generated by simulating the synthesized circuit at the "known" 64 corners; in the actual synthesis run, the worst case corners are dynamically chosen, hence only a fraction of these corners are actually considered (as we will detail in the coming paragraphs). The nominal synthesis design of Chapter 6 can have a random offset voltage of up to 4mV. The non-linear infinite programming formulation of the analog circuit synthesis problem was set up with a random offset voltage specification of 2mV. For clarity, in the graph, we have sorted the corners by the size of the random offset voltage in the nominal ASTRX/OBLX design. From the graph it is clear that about half of the 64 corners need to be considered by the formulation. It is also clear that the optimization has reduced the random offset voltage only as much as was needed to meet the specification. For the half of the worst-case corners that are not active constraints, the NLIP optimization returns a design whose random offsets are actually greater than that of the nominal design for that corner. While these corners are currently inactive, at a different sizing and biasing for the circuit, those corners can easily become active. This means it is

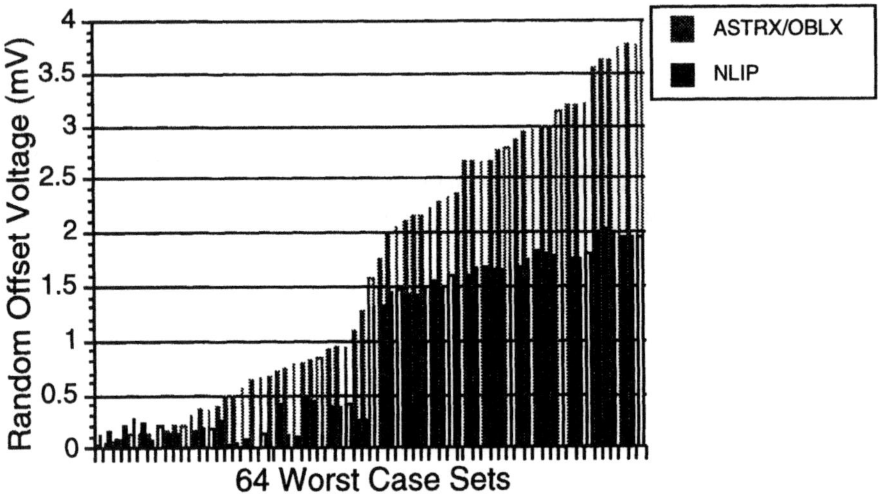

FIGURE 11.7 OTA Random Offset Voltage

impossible for the designer using a nominal tool to *a priori* define the minimal list of worst-case points. In an incompletely specified problem, the optimization will find a solution that might violate a constraint that was not specified.

Our approach can dynamically determine the actual worst-case points, using the inner annealing and the large-scale sensitivity of the cost function terms to the worst-case variables, hence it limits the designer's responsibility to just providing the operating point ranges. As expected, the device sizes from NLIP are larger to reduce sensitivity to mismatch. This is shown in Table 11.2 comparing the nominal and the NLIP results, with the execution times of both runs. Note that the execution time of this run for this prototype implementation is only 8 times more than the nominal design. The total number of critical points determined during this run was 12, thus the outer optimization was evaluating 12 circuits simultaneously for each move it made during the end of the synthesis. In addition, since the search space is more complicated, the annealer for the outer optimization has to do more moves to efficiently search the search space. The sequential addition of the circuits accounts for the less than 12x run-time degradation when the 12 critical points are identified.

A brute-force attack would just exhaustively search all the 64 "known" corners, while a designer-specified approach might do less than the 12 we chose dynamically. The former case is unnecessarily conservative, while the latter inevitably will consider the

TABLE 11.2 Device sizes (in microns) of the OTA and parametric manufacturing variations

Design	ASTRX/OBLX (nominal)	ASTRX/OBLX NLIP
W_{M1}	26	71
L_{M1}	2.7	3.3
W_{M3}	4.8	15
L_{M3}	4.6	8.6
W_{M5}	6.6	58
L_{M5}	2.1	15
CPU Time on RS6000/58H	10 min	80 min

wrong corners since the designer has no *a priori* knowledge of the sequence of designs the synthesis tool will consider.

Figure 11.8 shows the effect of the inclusion of the simplified manufacturing variations into the synthesis process. We use the same metric used in Chapter 8 for this comparison, namely, post-design yield optimization. More precisely, we start off with two nominal designs: one synthesized by the nominal version of ASTRX/OBLX and the second synthesized by the NLIP formulation. We use yield optimization to improve the yield of both these circuits. Instead of the traditional yield optimization algorithms, we pursue an indirect method detailed in [Krishna 95].The actual yield optimization approach used was a penalty function method that penalizes designs that do not meet the desired circuit performances, while at the same time trying to improve the circuit's immunity to manufacturing line variations. At every iteration in this gradient-based approach we evaluate the yield using a Monte-Carlo analysis, and this yield is plotted on the graph in Figure 11.8. The non-linear infinite programming formulation provides a good starting point for the yield optimization, much like the manual design did in our earlier example. As in the previous case, the nominally synthesized design is a difficult starting point for yield optimization. The run with the nominal synthesis starting point terminates due to an inability to find a direction of improvement.

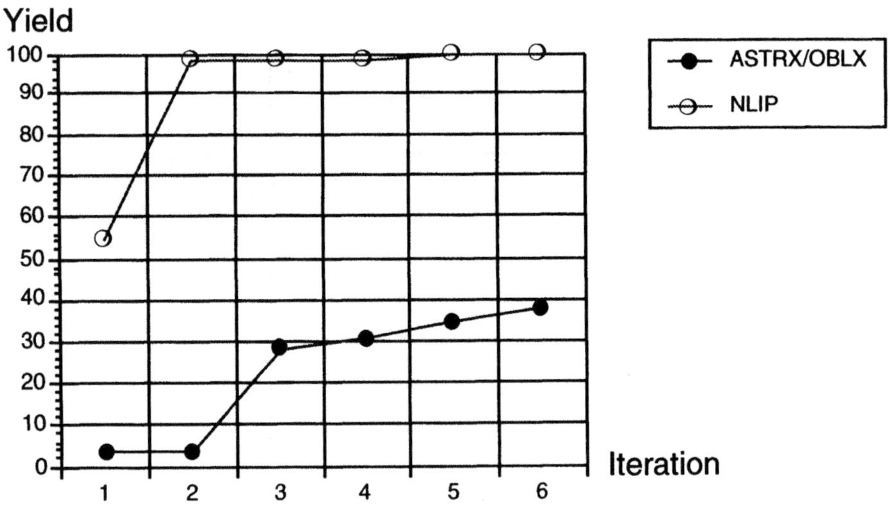

FIGURE 11.8 Post-design Yield Optimization

11.4 Band-Gap Reference Circuit

We have seen in our previous experiments that the variation-tolerant formulation can synthesize more manufacturable op-amps. In this section we present a completely novel synthesis result. Stable voltage and current references are required in practically every analog circuit. A common approach to providing such a reference is the Band-Gap reference circuit, but band-gap reference circuits are some of the most difficult circuits to design. In industry, precision band-gap circuits are still solely designed by highly experienced expert designers. The circuit we have chosen for this experiment is a simple first-order band-gap reference circuit.

Figure 11.9 shows the schematic of the band-gap reference circuit we synthesized for this experiment. It is based on the CMOS voltage reference described in [Tham 95], but has been modified to generate a bias current instead.

To understand the temperature dependence of this circuit, let us look at the Kirchhoff's Voltage Law equation in Equation 11.1, which describes the op-amp feedback

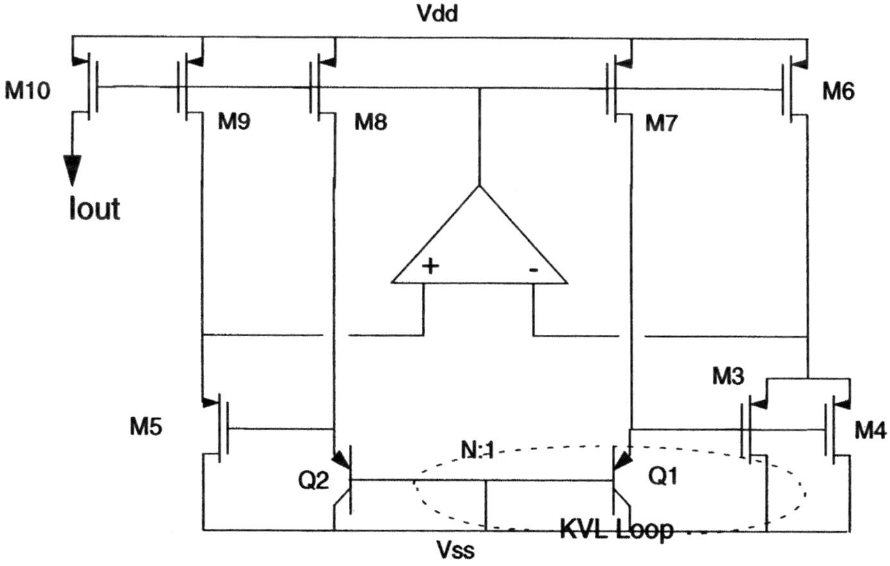

FIGURE 11.9 Simple Band-gap Reference Circuit

loop indicated by the dotted loop in Figure 11.9, with $V_{GS3,4}$ being the gate-source voltage for both the M3 and M4 devices.

$$V_{BE2} + V_{GS5} - V_{GS3,4} - V_{BE1} = 0 \tag{11.1}$$

In addition, recall that for a bipolar junction transistor, we can write an exponential relationship between the base-emitter voltage and collector current as Equation 11.2.

$$V_{BE} = V_T \cdot \ln\left(\frac{I}{I_s}\right) \tag{11.2}$$

For a MOS transistor, the relationship for a gate-source voltage as a function of the drain current follows a square law as given in Equation 11.3(with the threshold voltage as V_t to distinguish from $V_T = kT/q$).

$$V_{GS} = V_t + \sqrt{\frac{IL}{k'W}} \tag{11.3}$$

In our implementation we force the ratioing of device emitters so that Q1 to Q2 is 1:20, and the device width and lengths so that M3, M4 and M5 are identical, that is, M3 and M4 share the same current that M5 sees. Therefore we can write Equation 11.4 and Equation 11.5:

$$V_T \cdot \ln\left(\frac{I}{I_{S2}}\right) + V_t + \sqrt{\frac{IL}{k'W}} - \left(V_t + \sqrt{\frac{IL}{2k'W}}\right) - V_T \cdot \ln\left(\frac{I}{I_{S1}}\right) = 0 \tag{11.4}$$

$$V_T \cdot \ln\left(\frac{I_{S1}}{I_{S2}}\right) + \left(1 - \frac{1}{\sqrt{2}}\right) \cdot \sqrt{\frac{IL}{k'W}} = 0 \tag{11.5}$$

These can be simplified to create Equation 11.6.

$$-3 \cdot V_T + 0.3 \cdot \sqrt{\frac{IL}{k'W}} = 0 \tag{11.6}$$

Now, solving for the current I, which is the same as I_{out}, we get Equation 11.7:

$$I = 100 \cdot (V_T)^2 \cdot k'\frac{W}{L} \tag{11.7}$$

In this equation $V_T = kT/q$, and $k' = \mu_p C_{ox}$ where $\mu_p \sim T^{-1.5}$, thus the output current varies as $I_{out} \sim T^{-0.5}$, which is the best we will be able to achieve if we synthesize this circuit.

In our band-gap synthesis experiment, we began with the schematic shown in Figure 11.9. While the schematic shows hierarchy, with an op-amp to control the feedback, we flattened the entire schematic in our input to the synthesis system, replacing the op-amp with the simple OTA shown in Figure 11.1. The matching considerations described above were used to determine which subset of the device geometries (MOS width and length and bipolar emitter areas) were the design variables. The output current was specified to be 25µA, and a test-jig was placed around the op-amp devices to code gain, unity gain frequency and phase margin specifications to ensure those devices behaved like an op-amp. It took about 10 minutes to synthesize a circuit to these specifications on a PowerPC 604 based machine.

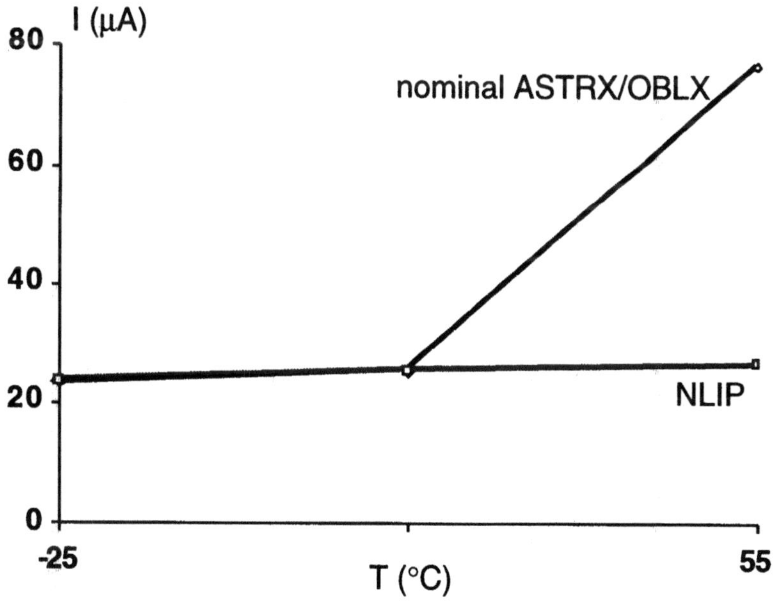

FIGURE 11.10 Temperature Sensitivity of Band-gap Reference

Figure 11.10 shows the graph of the output current with respect to temperature. As can be seen from the ASTRX/OBLX design, a nominal synthesis of this circuit is extremely unstable with respect to temperature. In comparison, the curve for the circuit from NLIP has a gentle slope with respect to temperature. As we have seen from the analysis above, this slope is to be expected (curve-fitting the data shows that the exponent is -0.42). We would need to expand the circuit topology to include curvature compensation resistors to ensure that dI_{out}/dT has zero slope at room temperature.

11.5 Summary

In this chapter we presented experimental results to verify the variation-tolerant non-linear infinite programming formulation discussed in Chapter 10. We implemented this formulation as an extension of the nominal approach used in ASTRX/OBLX to create ASTRX/OBLX NLIP. We then synthesized several circuits, including several involving operating point variations and simplified manufacturing line variations. These are the first synthesis results of circuits that can actually be used by a mixed-signal ASIC designer. We have shown this by comparing designs from the NLIP formulation with nominal synthesis results. In addition we have presented the first synthesis of a band-gap reference circuit. These circuits by their very nature cannot be synthesized by nominal synthesis alone. They require synthesis in the face of operating point variations, which we can now do with the variation-tolerant non-linear infinite programming formulation. In the next chapter we enumerate the contributions of our work and summarize directions for future work.

CHAPTER 12

Conclusions and Future Work

12.1 Introduction

In the previous chapters we have described a new approach to fully automated circuit design for high-performance analog cells and its implementation as the ASTRX/OBLX synthesis system. Our presentation was organized into two major parts. In the first part, from Chapter 2 to Chapter 7, we focused on the shortcomings in accuracy, automation, and breadth of previous approaches to analog synthesis, creating a new synthesis approach that we implemented as the tools ASTRX/OBLX. In the second part, we built upon the foundation of the ASTRX/OBLX nominal synthesis system and extended it to handle operating point and manufacturing line variations, creating a truly practical analog synthesis system. In this chapter, we highlight the contributions of this work and outline some areas for future investigation.

12.2 Contributions

In this section we enumerate the principal contributions of our work in analog circuit synthesis at the cell level. These contributions are:

- **An optimization formulation for analog synthesis that uses AWE for performance evaluation.** Ours is the first analog synthesis system that employs AWE as the core circuit performance evaluation technology. Because AWE is extremely efficient without sacrificing accuracy, this innovation allows the exploration of a substantially larger set of designable circuits. Without AWE, simulated annealing would take impracticably long to optimize realistic circuits.

- **A relaxed-D.C. circuit formulation within an unconstrained optimization framework.** Although others initially formulated Kirchhoff's laws as part of the constraints on an analog synthesis problem [Maulik 92a, Lipparini 82], they then solved it using a constrained optimization technique. In contrast, because simulated annealing is an unconstrained optimization technique, we further relax the D.C. constraints, allowing the dynamic weight control mechanism to control the D.C. terms of the cost function. Thus, the optimizer must converge to a solution that simultaneously satisfies both Kirchhoff's laws and meets user-specified performance specifications.
- **The use of fully encapsulated device models.** Ours is the first synthesis system that completely encapsulates device modelling information within device specific modules. In this manner, we gain the full detail of models used within detailed circuit simulators and are completely independent of the kind of device and the modelling equations. For example, it is a simple step to incorporate bipolar devices within ASTRX/OBLX and requires neither architectural nor code changes.
- **A Two-Phase synthesis architecture that pre-compiles the synthesis problem description then links it with the solution engine.** Ours is the first system that employs a separate analysis program (ASTRX) that generates executable code designed specifically to solve the synthesis problem at hand. This is a much more flexible strategy than previous approaches and allows unprecedented automation of the synthesis problem.
- **A new combination of annealing schedule and move control mechanism.** We have removed the simple move contraction mechanism for the modified Lam cooling schedule [Lam 88] and replaced it with a modified version of the dynamic move class selection mechanisms attributed to Hustin [Hustin 87]. This combination of the best characteristics of two annealing control mechanisms allows us to utilize a very complex move-set with the provably efficient cooling techniques of Lam.
- **A dynamic weight control mechanism.** We have automated the task of optimizing a cost function that may include several hundred scalar weights intended to balance competing terms and assure that other terms are forced to zero. We have provided problem independent trajectories that each term should follow, automatic tuning of these trajectories, and a feedback control mechanism that adjusts the weights such that the trajectories are followed. This strategy has created an optimizer that is robust over different circuits and specifications.

- **An optimization-based mechanism for tuning annealing parameters.** We have designed a system that adjusts the annealer's control parameters automatically. This system uses a local optimization technique to systematically tune these parameters such that we obtain consistently good results across various different synthesis problems. By tuning these parameters, we have freed the analog circuit designer from the arduous task of adjusting the annealer's controls, which might otherwise be required to perform a circuit synthesis task.
- **Demonstration of useful functionality.** Experiments with ASTRX/OBLX have demonstrated its ability to design circuits whose nominal performance exceeds that of manually tuned circuits. Further, in comparing with prior approaches, we have demonstrated speedups of as much as three orders of magnitude for a first time design problem coupled with improved performance prediction accuracy. We have also demonstrated ASTRX/OBLX is scalable and general by designing the cells for a high-performance pipelined A/D converter. Finally, experience by other users has given us confidence that the tools can be used by others.
- **A Non-Linear Infinite Programming formulation of the analog circuit design problem.** This variation-tolerant formulation allows for more complete modelling of the important design issues necessary to design practical cell-level analog circuits. This formulation also allows us to unify two characteristics of analog circuit design that were to date de-emphasized by nominal synthesis approaches: circuit environment variations and simplified fabrication line variations formulated in the style of parameter tolerances. We can now treat both of these supra-nominal synthesis concepts in identically the same formulation without loss of the meaning behind these concepts. Thus, even though we do "force" manufacturing line variations into this formulation, we are able to use the formulation to design cells that are less sensitive to the manufacturing line variations we have incorporated, and, more importantly, that have an initial yield high enough for subsequent manual or automatic yield optimization to be feasible.
- **An annealing within annealing implementation of the Non-Linear Infinite Programming formulation.** By implementing our Non-Linear Infinite Programming algorithm as an extension to the ASTRX/OBLX nominal synthesis framework, we are able to leverage several of the important features of ASTRX/OBLX's successful nominal synthesis strategy.
- **Demonstration of manufacturable synthesis.** We have shown results of experiments demonstrating the ability to design more manufacturable analog cells. Further, in comparing with previous nominal synthesis approaches, we have highlighted more clearly the limitations of nominal synthesis strategies, and have demonstrated that we can design circuits that arc more practical for use in mixed-signal applications.

- **Synthesis results for a new class of analog cells.** We have extended the suite of automatically synthesized circuits to a new class of analog cells: the band-gap reference circuit.

12.3 Future Work

Both the nominal and the NLIP versions of ASTRX/OBLX have been shown to be effective systems for the automatic design of analog cells. However, this success in itself opens up many avenues of possible further research. In this section, we describe several interesting areas of ongoing and future work.

- **Post-synthesis tuning.** Simulated annealing has been shown to be effective at finding global or near global minima of complex cost surfaces where many minima exist. Despite this, in the immediate region of the minima annealing is not very efficient and takes a substantial amount of time to find the absolute local minimum. To combat this problem, Stanisic [Stanisic 96] has experimented with using local, gradient-based optimization techniques when the annealing process is complete or nearly complete. Although simulated annealing alone found better minima than just the gradient-based method, simulated annealing *followed* by gradient-based optimization consistently yielded the best results. This suggests a similar strategy could benefit ASTRX/OBLX. There are several hurdles to overcome here. First, the gradient-based approaches require either numeric or symbolic gradients, which will require substantial additional work within either OBLX, ASTRX, or both. Further, gradient-based techniques need significant effort to ensure that they are numerically stable, which will likely entail scaling or tuning of the cost function or independent variables. Finally, the most appropriate point to switch from annealing to the gradient-based approach must be determined. Clearly, switching too early would risk finding only poor minima, while waiting until after the anneal is complete might greatly increase run-time. For example, in [Stanisic 96], Stanisic found that starting with a solution found by annealing and performing a gradient-based optimization with numerically determined gradients took about 50% of the run-time of the annealing itself. Thus, the overall run-time was about 50% longer. However, some of this increase could be mitigated by stopping the annealing sooner.
- **Topology design.** Recall that our synthesis methodology does not address the topological aspects of circuit design. There are several techniques that could be used to incorporate topological design into the ASTRX/OBLXsystem. The most straightforward of these would be to use a topology selection expert system as a

pre-processing step. Based on user-supplied specifications, this expert system could select a set of appropriate topologies for detailed design and pass them on to ASTRX/OBLX. A somewhat more interesting alternative would be to evolve the topology, sizing, biasing and manufacturing constraints simultaneously. This would require substantial additions to the OBLX move-set and presents some interesting issues involving the relaxed D.C. formulation. Because different circuits would have substantially different operating points, some technique would need to be used to compensate for the change in C^{KCL} when the topology is altered. Exploratory work on this extension was performed with encouraging results [Jones 96].

- **Homotopy and other D.C. solution methods.** It is difficult to find a D.C. operating point for certain circuits. Researchers in simulation are starting to apply strong global optimization techniques, such as homotopy [Richter 83] via the application interfaces of HSPICE [Meta 87] or SABRE [Getreu 89], rather than Newton-Raphson to the operating point problem. An interesting experiment would be to replace the Newton-Raphson solver used as part of the OBLX move-set with one of these more reliable techniques. Although OBLX does not force Kirchhoff's laws to be met at all times, it is crucial that it can force a D.C. solution when it needs to. This might further improve the robustness of the system.
- **Exploitation of inherent parallelism in outer optimization.** The largest single contributor to the growth of the synthesis run-times in our NLIP results has been the need for designing the same circuit to face multiple environmental or manufacturing process critical points generated from the inner optimizations. Each of these circuits can be simulated on a separate computer, with the vector of actual circuit performances sent back to a central computer to tally their contributions to the overall cost and, to apply the annealing algorithm to guide the search process.
- **Application to Micro-Electro-Mechanical system synthesis.** Micro-Electro-Mechanical systems are composed of small numbers of both electronic and mechanical elements for which linear and equation-based models can be written. In addition, range variables are critical in the design of such systems, which have to be stable in the specified range of environment parameters where they will operate. Therefore the approach presented in this book can easily be applied to this application area to automate design of MEMS systems. See [Fedder 96] for more of our initial ideas in this direction.
- **Revisiting simulation-based synthesis.** Rapid progress in machine speed is making "SPICE-in-the-loop" synthesis increasingly attractive. Krasnicki at Carnegie Mellon [Krasnicki 97] has combined a new annealer in the style of OBLX with a genetic algorithms component and also distributed the computation over a net-

work of up to 35 workstations in parallel. Early results are encouraging and use a full circuit simulator (Cadence Spectre) in the loop. The ASTRX/OBLX optimization-based synthesis architecture was the starting point for this very recent work.

Bibliography

[Aaron 56] M. R. Aaron. The Use of Least Squares in System Design. *IRE Transactions on Circuit Theory.* CT-3224-231, December, 1956.

[Aarts 87] E. H. Aarts and P. J. van Laarhoven. *Simulated Annealing: Theory and Applications.* Reidel, Dordrecht, 87.

[Aho 87] A. V. Aho, R. Sethi, and J.D. Ullman. *Compilers: Principles, Techniques, and Tools.* Addison-Wesley Publishing Company, Reading, Massachusetts, 1987.

[Alaybeyi 90] M. M. Alaybeyi. *BUSTLE: A New Circuit Simulation Tool.* Master's thesis, Bilkent University, Turkey, July, 1990.

[Allen 83] Allen P. E. A Design Methodology for CMOS Analog Integrated Circuits. *Proc. of IEEE International Conference on CAD.* 1983.

[Astap 73] IBM. *The Advanced Statistical Analysis Program (ASTAP)* Armonk, NY, 1973.

[Bandler 78] J. W. Bandler and H. L. Abdel-Malek. Optimal Centering, Tolerancing, and Yield Determination via Updated Approximations and Cuts. *IEEE Transactions on Circuits and Systems.* CAS-25(10):853-871, October, 1978.

[Basaran 93] B. Basaran, R. A. Rutenbar and L. R. Carley. Latchup-Aware Placement and Parasitic-Bounded Routing of Custom Analog Cells. *Proceedings of the IEEE/ACM International Conference on Computer-Aided Design.* 1993.

[Bassak 95] G. Bassak, "Analog and Mixed-Signal Cells", *Integrated System Design,* Vol. 7, No. 74, August 1995, pp. 36-54.

[Bazaraa 79] M. S. Bazaraa and C. M. Shetty. *Non-Linear Programming: Theory and Applications.* Wiley, 1979.

[Berkcan 88] E. Berkcan, M. d'Abreu, and W. Laughton. Analog Compilation Based On Successive Decompositions. *25thACM/IEEE Design Automation Conference.* 369-375, 88.

[Bhatti 79] M. A. Bhatti, K. S. Pister, and E. Polak. Optimal Design of an Earthquake Isolation System. *Proc. IUTAM Symp. on Structural Control, Univ. of Waterloo, Waterloo, Ont. Canada.* June, 1979.

[Bilbro 91] G. L. Bilbro and W. E. Snyder. Optimization of Functions with Many Minima. *IEEE Trans. on Systems, Man and Cybernetics.* 21(4):840-849, July/August, 1991.

[Bonomi 84] E. Bonomi and J. Dutton. The N-City Travelling Salesman Problem: Statistical Mechanics and the Metropolis Algorithm. *SIAM Review.* 26(4):551-568, October, 1984.

[Bowman 85] Bowman Robert J. and Lane David J. A Knowledge-Based System for Analog Integrated Circuit Design. *IEEE International Conference on Computer-Aided Design.* 210-212, 1985.

[Brayton 80] R. K. Brayton, S. W. Director and G. D. Hachtel. Yield Maximization and Worst-Case Design with Arbitrary Statistical Distributions. *IEEE Transactions on Circuits and Systems.* CAS-27(9):756-764, September, 1980.

[Brayton 80a] R. K. Brayton and R. Spence. *Sensitivity and Optimization.* Elsevier, New York, 1980.

[Brayton 81] R. K. Brayton, G. D. Hachtel, and A. L. Sangiovanni-Vincentelli. A Survey of Optimization Techniques for Integrated-Circuit Design. *Proceedings of the IEEE.* 69(10):1334-1362, October, 1981.

[Brayton 90] R. K. Brayton, A. L. Sangiovanni-Vincentelli, and G. D. Hachtel. Multi-level Logic Synthesis. *Proceedings of IEEE.* 78(2), February, 1990.

[Brent 73] R. P. Brent. *Algorithms for Minimization Without Derivatives.* Prentice-Hall, Englewood Cliffs, NJ, 1973.

[Brown 85] David C. Brown. Capturing Mechanical Design Knowledge. *ASME CIME.* Boston, August, 1985.

[Burr 90] J. G. Graeme, G. E. Tobey, L. P. Huelsman (editor). *Operational Amplifiers: Design and Applications.* McGraw-Hill Book Company, 1990.

[Carley 93] L. R. Carley, P. C. Maulik, E. S. Ochotta, and R. A. Rutenbar. Analog Cell-Level Synthesis Using a Novel Problem Formulation. *Advances in Analog Circuit Design.* In J. H. Huijsing, Kluwer Academic Press, 1993.

[Carley 96] L. R. Carley, G. E. Gielen, R. A. Rutenbar, and W. C. Sansen. Synthesis Tools for Mixed Signal ICs: Progress on Frontend and Backend Strategies. *Proc. 1996 ACM/IEEE Design Automation Conference*, pages 298-303. June, 1996.

[Castello 85] R. Castello and P. R. Gray. Performance Limitations in Switched-Capacitor Filters. *IEEE Trans. on Circuits and Systems.* CAS-32(9), September, 1985.

[Chan 91] P. K. Chan. Comments on Asymptotic Waveform Evaluation for Timing Analysis. *IEEE Transactions on Computer-Aided Design.* 10(8):1078-79, August, 1991.

[Collet 91] R. Collet and J. DeArmon. Interpreting the Mixed Signals of the Mixed-Signal Market. Dataquest, November 1991.

[Cohen 88] Brian Cohen, Edmond Cooley, et al. *Analog Hardware Description Language: A Description.* Technical Report, Thayer School of Engineering, Dartmouth College, July, 1988.

[Cohn 91] J. M. Cohn, D. J. Garrod, R. A. Rutenbar and L. R. Carley. KOAN/ANAGRAM II: New Tools for Device-Level Analog Placement and Routing. *IEEE J. of Solid-State Circuits.* 26(3):330-342, March, 1991.

[Cohn 92] J. M. Cohn. *Automatic Device Placement for Analog Cells in KOAN.* Ph.D. thesis, Carnegie Mellon University, February, 1992.

[Coope 85] I. D. Coope and G. A. Watson. A Projected Lagrangian Algorithm for Semi-Infinite Programming. *Mathematical Programming.* 32337-356, 1985.

[Cormen 90] T. H. Cormen, C. E. Leiserson and R. L. Rivest. *Introduction to Algorithms.* The MIT Press, McGraw-Hill Book Company, 1990.

[Cox 85] P. Cox, P. Yang, S. S. Mahant-Shetti and P. Chatterjee. Statistical Modeling for Efficient Parametric Yield Estimation of MOS VLSI Circuits. *IEEE Trans. on Electron Devices.* ED-32(2):471-478, February, 1985.

[Davis 82] Davis. Randall, and Douglas B. Lenat. *Knowledge-Based Systems in Artificial Intelligence.* McGraw-Hill International book Company, 1982.

[Degrauwe 84] M. G. R. DeGrauwe and W. M. C. Sansen. The Current Efficiency of MOS Transconductance Amplifiers. *IEEE Journal of Solid-State Circuits.* SC-19(3), June, 1984.

[Degrauwe 84a] M. DeGrauwe and W. M. Sansen. A Synthesis Program for Operational Amplifiers. *International Solid-State Circuit Conference*, pages 18-19. IEEE, 1984.

[Degrauwe 87] M. G. R. DeGrauwe, O. Nys, E. Dijkstra, J. Rijmenants, S. Bitz, B. L. A. G. Goffart, E. A. Vittoz, S. Cserveny, C. Meixenberger, G. van der Stappen and H. J. Oguey. IDAC: An Interactive Design Tool for Analog CMOS Circuits. *IEEE Journal of Solid-State Circuits.* SC-22(6):1106-1116, December, 1987.

[Degrauwe 87a] M. G. DeGrauwe, et. al. An Analog Expert Design System. *IEEE International Solid-State Circuits Conference.* 1987.

[Degrauwe 89] M. G. R. DeGrauwe, B. L. A. G. Goffart, C. Meixenberger, M. L. A. Pierre, J. B. Litsios, J. Rijmenants, O. J. A. P. Nys, E. Dijkstra, B. Joss, M. K. C. M. Meyvaert, T. R. Schwarz, and M. D. Pardoen. Towards an Analog System Design Environment. *IEEE Journal of Solid-State Circuits.* SC-24(3):659-672, June, 1989.

[Dekleer 80] J. De Kleer, and G. J. Sussman. Propagation of Constraints Applied to Circuit Synthesis. *Circuit Theory and Applications.* 8 1980.

[Dekleer 85] J. de Kleer. How Circuits Work. *Qualitative Reasoning About Physical Systems.* In D. G. Bobrow, The MIT Press, 1985, pages 205-280.

[Director 69] S. W. Director and R. A. Rohrer. Automated Network Design - The Frequency-Domain Case. *IEEE Trans. on Circuit Theory.* CT-16(5):330-337, August, 1969.

[Director 69a] S. W. Director and R. A. Rohrer. The Generalized Adjoint Network and Network Sensitivities. *IEEE Trans. on Circuit Theory.* CT-16(3):318-323, August, 1969.

[Director 77] S. W. Director and G. D. Hachtel. The Simplicial Approximation Approach to Design Centering. *IEEE Transactions on Circuits and Systems.* CAS-24(7):363-372, July, 1977.

[Director 92] S. W. Director, P. Feldmann and K. Krishna. Optimization of Parametric Yield: A Tutorial. *IEEE Custom Integrated Circuit Conference*, pages 3.1/1-8. May, 1992.

[Dixon 78] L. C. W. Dixon and G. P. Szego. *Towards Global Optimisation 2.* North-Holland, 1978.

[Doernberg 84] J. Doernberg, H.-S. Lee, and D. A. Hodges. Full-Speed Testing of A/D Converters. *IEEE Journal of Solid-State Circuits.* SC-19(6):820-827, December, 1984.

[Eaves 71] B. C. Eaves and W. I. Zangwill. Generalized Cutting Plane Algorithms. *SIAM Journal of Control.* 9529-542, 1971.

[Ecap 65] IBM. *1620 Electronic Circuit Analysis Program* 1965.

[Elgamal 84] A. El Gamal and I Shperling. Design of Good Codes via Simulated Annealing. *Workshop on Statistical Physics in Engineering and Biology.* IBM T. J. Watson Research Center, April, 1984.

[Elturky 86] F. M. El-Turky and R. A. Nordin. BLADES: An Expert System for Analog Circuit Design. *IEEE International Symposium on Circuits and Systems.* 1986.

[Elturky 89] F. El-Turky and E. E. Perry. BLADES: An Artificial Intelligence Approach to Analog Circuit Design. *IEEE Transactions on Computer-Aided Design.* 8(6):680-692, June, 1989.

[Fedder 96] G. K. Fedder, S. Iyer, and T. Mukherjee. "Automated Optimal Synthesis of Microresonators," Proc. Transducers '97, Chicago, IL, June 1997, pp. 1109-1112.

[Feldmann 93] P. Feldmann and S. W. Director. Integrated Circuit Quality Optimization using Surface Integrals. *IEEE Transactions on Computer-Aided Design.* 12(12):1868-1879, December, 1993.

[Fernandez 92] F. V. Fernandez et al. Accurate Simplification of Large Symbolic Formulae. *Proceedings of IEEE International Conference on Computer-Aided Design*, pages 318-321. November, 1992.

[Fiacco 83] A. V. Fiacco and K. O. Kortanek, eds. *Semi-Infinite Programming and Applications, Lecture notes in Economics and Mathematical Systems 215.* Springer-Verlag, 1983.

[Fleisher 85] H. Fleisher, J. Giraldi, D. B. Martin, R. L. Phoenix, and M. A. Tavel. Simulated Annealing as a Tool for Login Optimization in a CAD Environment. *Proceedings of the IEEE/ACM International Conference on Computer-Aided Design*, pages 203-205. 1985.

[Fletcher 90] R. Fletcher. *Practical Methods of Optimization.* Wiley, 1990.

[Fung 88] A. H. Fung, D. J. Chen, Y. N. Lai, B. J. Sheu. Knowledge-Based Analog Circuit Synthesis with Flexible Architecture. *IEEE International Conference On Computer Design.* 48-51, October, 88.

[Garey 79] M. R. Garey and D. S. Johnson. *Computers and Intractability: A Guide to the Theory of NP-Completeness.* Freeman, San Francisco, CA, 1979.

[Garrod 88] D. J. Garrod, R. A. Rutenbar, and L. R. Carley. Automatic Layout of Custom Analog Cells in ANAGRAM. *IEEE International Conference on Computer-Aided Design*, pages 544-547. November, 1988.

[Gelfand 91] S. B. Gelfand and S. K. Mitter. Simulated Annealing Type Algorithms for Multivariate Optimization. *Algorithmica.* 6419-436, 1991.

[Getreu 89] I. Getreu. Behavioral Modelling of Analog Blocks using the SABER Simulator. *Proc. Microwave Circuits and Systems.* 977-980, August, 1989.

[Gielen 89] G. E. Gielen, H. C. C. Walscharts, and W. M. C. Sansen. ISAAC: A Symbolic Simulator for Analog Circuits. *IEEE Journal of Solid-State Circuits.* SC-24(6):1587-1597, December, 1989.

[Gielen 90] G. E. Gielen, H. C. Walscharts, and W. C. Sansen. Analog Circuit Design Optimization Based on Symbolic Simulation and Simulated Annealing. *IEEE Jour. of Solid State Circuits.* SC-25(3):707-713, June, 1990.

[Gielen 90a] G. E. Gielen. *Design Automation for Analogue Integrated Circuits.* Ph.D. thesis, Katholieke Universiteit Leuven, October, 1990.

[Gielen 91] G. Gielen and W. Sansen. *Symbolic Analysis for Automated Design of Analog Integrated Circuits.* Kluwer, 1991.

[Gielen 92] G. Gielen. Private Communication. 1992.

[Gill 81] P. E Gill, W. Murray, and M. H. Wright. *Practical Optimization.* Academic Press, 1981.

[Goldberg 89] D. E. Goldberg. *Genetic Algorithms in Search, Optimization and Machine Learning.* Addison-Wesley, Massachusetts, 1989.

[Gray 84] P. R. Gray and R. G. Meyer. *Analysis and Design of Analog Integrated Circuits.* John Wiley & Sons, 1984.

[Gray 87] P. G. Gray. Analog IC's in the Submicron Era: Trends and Perspectives. *IEEE IEDM Digest of Technical Papers.* 5-9, 1987.

[Gray 95] P. R. Gray. Possible Analog IC Scenarios for the 90's. *VLSI Symposium.* 1995.

[Grossmann 91] I. E. Grossmann and D. A. Straub. Recent Developments in the Evaluation and Optimization of Flexible Chemical Processes. *Proceedings of COPE-91*, pages 49-59. 1991.

[Guardiani 93] C. Guardiani and J. Benkoski. Flow-Driven Graphical User Interface For Parametric and Statistical Design Optimization. *IEEE Custom Integrated Circuit Conference*, pages 13.5/1-4. San Diego, May, 1993.

[Gummel 70] H. K. Gummel and H. C. Poon. An Integral Charge Control Model of Bipolar Transistors. *The Bell System Technical Journal.* 827-851, May, 1970.

[Hammersley 64] J. M. Hammersley and D. C. Handscomb. *Monte Carlo Methods.* Methuen, London, 1964.

[Harjani 87] R. Harjani, R. A. Rutenbar and L. R. Carley. A Prototype Framework for Knowledge-Based Analog Circuit Synthesis. *24th Design Automation Conference*, pages 42-49. ACM/IEEE, June, 1987.

[Harjani 88] R. Harjani, R. A. Rutenbar and L. R. Carley. Analog Circuit Synthesis for Performance in OASYS. *IEEE International Conference on Computer-Aided Design*. November, 1988.

[Harjani 89] R. Harjani, R. A. Rutenbar, and L. R. Carley. OASYS: A Framework for Analog Circuit Synthesis. *IEEE Transactions on Computer-Aided Design*. CAD-8(12):1247-1266, December, 1989.

[Harjani 89a] R. Harjani. *OASYS: a framework for analog circuit synthesis*. Ph.D. thesis, Carnegie Mellon University, March, 1989.

[Harvey 92] J. P. Harvey, M. I. Elmasry, B. Leung. STAIC: An Interactive Framework for Synthesizing CMOS and BiCMOS Analog Circuits. *IEEE Trans. on Computer-Aided Design*. CAD-11(11):1402-1418, November, 1992.

[Hashizume 89] M. Hashizume, H. Y. Kawai, K. Nii, T. Tameseda. Design Automation System for Analog Circuits Based on Fuzzy Logic. *IEEE Custom Integrated Circuit Conference*, pages 4.6/1-4. IEEE, San Diego, May, 1989.

[Hauser 86] M. W. Hauser and R. W. Brodersen. Circuit and Technology Considerations for MOS Delta-Sigma A/D Converters. *IEEE Intl. Symp. on Circuits and Systems*. May, 1986.

[Henderson 93] R. K. Henderson, L. Astler, A. El Khalifa, and M. Degrauwe. A Spreadsheet Interface for Analog Circuit Design Knowledge Capture and Re-use. *IEEE Custom Integrated Circuit Conference*, pages 13.3/1-4. May, 1993.

[Hobekotte 87] E. Hobekotte, B. Hoefflinger, H.-W. Klien, M. A. Beunder. State of the Art in the Analog CMOS Circuit Design. *Proceedings of IEEE*. 75(6):816-828, June, 1987.

[Hocevar 85] D. E. Hocevar, P. Yang, T. N. Trick, and B. D. Epler. Transient sensitivity computation for MOSFET circuits. *IEEE Transactions on Computer-Aided Design.* CAD-4(4):609-620, October, 1985.

[Hocevar 88] D. E. Hocevar, P. F. Cox, and P. Yang. Parametric Yield Optimization for MOS circuit blocks. *IEEE Transactions on Computer-Aided Design.* CAD-7(6):645-658, June, 1988.

[Hocevar 90] D. E. Hocevar, R. Arora, U. Dasgupta, S. Dasgupta, N. Subramanyam, and S. Kashyap. A Usable Circuit Optimizer for Designers. *Proceedings of the IEEE/ACM International Conference on Computer-Aided Design*, pages 290-293. November, 1990.

[Holland 92] J. Holland. Genetic Algorithms. *Scientific American.* 7(7):66-72, July, 1992.

[Horst 90] R. Horst. Deterministic methods in constrained global optimization: some recent advances and new field of application. *Naval Research Logistics.* 37(4), August, 1990.

[Horst 93] R. Horst. *Global Optimization: Deterministic Approaches.* Springer-Verlag, Berlin, 1993.

[Huang 90] X. Huang. *Pade Approximation of Linear(ized) Circuit Responses.* Ph.D. thesis, Carnegie-Mellon University, November, 1990.

[Huang 86] M. D. Huang, F. Romeo and A. Sangiovanni-Vincentelli. An Efficient General Cooling Schedule for Simulated Annealing. *Proceedings of International Conference on Computer-Aided Design*, pages 381-384. November, 1986.

[Huhns 88] M. N. Huhns and R. D. Acosta. Argo: A System for Design by Analogy. *IEEE Expert.* 3(3):53-68, 1988.

[Hustin 87] S. Hustin and A. Sangiovanni-Vincentelli. TIM, a new standard cell placement program based on the simulated annealing algorithm. *IEEE Physical Design Workshop on Placement and Floorplanning*. April, 1987.

[Hsu 94] J. Hsu and C Sechen. Fully Symbolic Analysis of Large Analog Integrated Circuits. *IEEE Custom Integrated Circuit Conference*, pages 457-464, May 1994.

[Intermetrics 88] Intermetrics, Inc. *MHDL Language Reference Manual.* Technical Report, ARPA/Tri-Service MHDL Initiative, July, 1988.

[Jepsen 84] D. W. Jepsen and C. D. Gelatt, Jr. Macro Placement by Monte Carlo Annealing. *Proceedings of the IEEE/ACM International Conference on Computer-Aided Design*, pages 495-498. November, 1984.

[Johnson 91] B. Johnson, T. Quarles, et al. *SPICE3 Version 3e User's Manual.* Technical Report, University of California, Berkeley, April, 1991.

[Jones 96] W. E. Jones, III. *Simultaneous Topology Selection and Sizing/Biasing for Analog Synthesis in ASTRX/OBLX.* Master's thesis, CMUCAD-96-03, Carnegie-Mellon University, January, 1996.

[Jongsma 91] J. Jongsma, et al. An Open Design Tool for Analog Circuits. *IEEE International Symposium on Circuits and Systems*, pages 2000-2003. June, 1991.

[Kalos 86] M. H. Kalos. *Monte Carlo Methods.* J. Wiley & Sons, New York, 1986.

[Kan 89] A. H. Rinnooy Kan, G. T. Timmer. Global Optimization. *Handbooks in Operations Research and Management Science, Vol. 1.* In G. L. Nemhauser, et al., Elsevier Science Publishers, B.V. (North-Holland), 1989, Chapter 9.

[King 85] P. King. Subcircuit on Linear Arrays - A New Array Topology. *IEEE Custom Integrated Circuit Conference.* 1985.

[Kirkpatrick 83] S. Kirkpatrick, C. D. Gelatt, and M. P. Vecchi. Optimization by Simulated Annealing. *Science.* 220(4598):671-680, November, 83.

[Klinke 92] R. Klinke, H. L. Fiedler, B. J. Hosticka, R. Kokozinski, and I. Munster. Rule-based Analog Circuit Design. *Proceedings of the European Design Automation Conference*, pages 480-4. IEEE, March, 1992.

[Knapp 86] Knapp, David W. and Parker, Alice C. A Design Utility Manager: the ADAM Planning Engine. *Proceedings of the IEEE/ACM Design Automation Conference*, pages 48-54. 1986.

[Koh 87] Koh H. Y., Sequin C. H. and Gray P. R. Automatic Synthesis of Operational Amplifiers Based on Analytic Circuit Models. *International Conference on Computer-Aided Design*, pages 502-505. IEEE, November, 1987.

[Koh 90] H. Y. Koh, C. H. Sequin and P. R. Gray. OPASYN: A Compiler for MOS Operational Amplifiers. *IEEE Trans. on Computer-Aided Design.* CAD-9(2), February, 1990.

[Krasnicki 97] K. .

[Krishna 95] K. Krishna and S. W. Director. The Linearized Performance Penalty (LPP) Method for Optimization of Parametric Yield and Its Reliability. *IEEE Transactions on Computer-Aided Design.* 14(12):1557-1568, Dec, 1995.

[Krishna 95a] K. Krishna and S. W. Director. A Novel Methodology for Statistical Parameter Extraction. *Proceedings of the IEEE/ACM International Conference on Computer-Aided Design*, pages 696-699. 1995.

[Kruiskamp 95] W. Kruiskamp and D. Leenaerts. DARWIN: CMOS Opamp Synthesis by Means of a Genetic Algorithm. *Proceedings of the IEEE/ACM Design Automation Conference*, pages 433-438. June, 1995.

[Kuh 90] E. S. Kuh and T. Ohtsuki. Recent Advances in VLSI Layout. *Proceedings of the IEEE.* 78(2):237-263, Feb, 1990.

[Kundert 89] K. S. Kundert and A. Sangiovanni-Vincentelli. Reply to: Comments on 'Simulation of Nonlinear Circuits in the Frequency Domain. *IEEE Transactions on Computer-Aided Design.* 8(8), August, 1989.

[Lam 86] J. Lam and J.-M. Delosme. Logic Minimization Using Simulated Annealing. *Proceedings of the IEEE/ACM International Conference on Computer-Aided Design*, pages 348-352. November, 1986.

[Lam 88] J. Lam. *An Efficient Simulated Annealing Schedule.* Ph.D. thesis, Yale University, Fall, 1988. Dept. of CS.

[Lam 88a] J. Lam and J. M. Delosme. Performance of a New Annealing Schedule. *Proceedings of IEEE/ACM Design Automation Conference*, pages 306-311. 1988.

[Lee 91] E. K. Lee and P. G. Gulak. A CMOS Field-Programmable Analog Array. *IEEE Journal of Solid-State Circuits.* sc-26(12):1860-1867, December, 1991.

[Leung 74] K. Leung and R. Spence. Efficient Statistical Circuit Analysis. *Electron Letters.* 10(17):377-378, 1974.

[Lightner 81] M. R. Lightner and S. W. Director. Multiple Criterion Optimization for the Design of Electronic Circuits. *IEEE Transactions on Circuits and Systems.* CAS-28(3):169-179, March, 1981.

[Lineback 87] J. R. Lineback. ASIC houses rush to add Analog Functions to Libraries. *Electronics.* 60(22):31, Oct 29, 1987.

[Lipparini 82] A. Lipparini, E. Marazzi, and V. Rizzoli. A New Approach to the Computer-Aided Design of Nonlinear Networks and its Application to Microwave Parametric Frequency Dividers. *IEEE Transactions on Microwave Theory Technology.* 301050-1058, July, 1982.

[Luenberger 84] D. G. Luenberger. *Linear and Nonlinear Programming.* Addison-Wesley, 1984.

[Madsen 79] K. Madsen and H. Schjaer-Jacobsen. Algorithms for Worst Case Tolerance Optimization. *IEEE Transactions on Circuits and Systems.* CAS-26September, 1979.

[Makris 92] C. A. Makris and C. Toumazou. Qualitative Reasoning in Analog IC Design Automation. *IEEE Custom Integrated Circuit Conference*, pages 8.3/1-4. IEEE, Boston, May, 1992.

[Makris 95] C. A. Makris and C. Toumazou. Analog IC Design Automation: Part II - Automated Circuit Correction by Qualitative Reasoning. *IEEE Transactions on Computer-Aided Design.* CAD-14(2):239-254, February, 1995.

[Maly 86] W. Maly, A. J. Strojwas and S. W. Director. VLSI Yield Prediction and Estimation: A Unified Framework. *IEEE Transactions on Computer-Aided Design.* CAD-5(1):114-130, January, 1986.

[Mathers 67] H. W. Mathers, S. R. Sedore, and J. R. Seuts. *Automated Digital Computer Program for Determining Responses of Electronic Circuits to Transient Nuclear Radiation.* IBM File 66-928-611, IBM Space Guidance Center, Oswego, NY, February, 1967.

[Maulik 91] P. C. Maulik and L. R. Carley. Automating Analog Circuit Design using Constrained Optimization Techniques. *Proceedings of IEEE/ACM International Conference on Computer-Aided Design*, pages 390-393. November, 1991.

[Maulik 92] P. C. Maulik, L. R. Carley and R. A. Rutenbar. A Mixed Integer Non-Linear Programming Approach to Analog Circuit Synthesis. *Proc. 29th ACM/IEEE Design Auto. Conf.*, pages 698-703. June 1992.

[Maulik 92a] P. C. Maulik. *Formulations for Optimization-based Synthesis of Analog Cells.* Ph.D. thesis, Carnegie Mellon University, September, 1992.

[Maulik 95] P. C. Maulik, L. R. Carley, and R. A. Rutenbar. Integer Programming Based Topology Selection of Cell-Level Analog Circuits. *IEEE Transactions on CAD.* CAD-14(4):401-412, April, 1995.

[Meta 87] Meta-Software. *HSPICE User's Manual.* Meta-Software Inc., Campbell, CA, 1987.

[Metropolis 53] N. Metropolis, A. W. Rosenbluth, M. N. Rosenbluth, and A. H. Teller. Equation of State Calculations by Fast Computer Machines. *J. Chem. Phys.* 21087, 1953.

[Mitra 92] S. Mitra, S. K. Nag, R. A. Rutenbar and L. R. Carley. System-level Routing of Mixed-Signal ASICs in WREN. *Proc. IEEE Intl. Conf. Computer-Aided Design.* 394-399, 1992.

[Nagel 71] L. W. Nagel and R. A. Rohrer. Computer Analysis of Nonlinear Circuits, Excluding Radiation (CANCER). *IEEE J. of Solid-State Circuits.* 6(4):166-182, August, 1971.

[Nagel 75] Nagel L. W. *SPICE2: A Computer Program to Simulate Semiconductor Circuits.* ERL Memo ERL-M520, University of California, May, 1975.

[Nakamura 92] K. Nakamura and L. R. Carley. An Enhanced Fully Differential Folded-Cascode Op Amp. *IEEE Journal of Solid-State Circuits.* SC-27(4):563-568, April, 1992.

[Nassif 84] S. R. Nassif, A. J. Strojwas and S. W. Director. FABRICS II: A statistically based IC fabrication process simulator. *IEEE Transactions on Computer-Aided Design.* CAD-3(1):20-46, January, 1984.

[Ning 91] Z.-Q. Ning, T. Mouthaan, H. Wallinga. SEAS: A Simulated Evolution Approach for Analog Circuit Synthesis. *IEEE Custom Integrated Circuit Conference*, pages 5.2.1-4. May, 1991.

[Nye 81] W. Nye, E. Polak, A. Sangiovanni-Vincentelli, A. Tits. DELIGHT: An Optimization-Based Computer-Aided Design System. *IEEE International Symposium on Circuits and Systems*, pages 851-855. April, 1981.

[Nye 84] W. Nye, et al. *DELIGHT.SPICE User's Guide* 1984.

[Nye 88] W. Nye, D. C. Riley, A. Sangiovanni-Vincentelli and A. L. Tits. DELIGHT.SPICE: An Optimization-Based System for Design of Integrated Circuits. *IEEE Trans. on Computer-Aided Design.* CAD-7(4):501-518, April, 1988.

[Nye 95] W. Nye. Private Communication. . 1995 .

[Ochotta 89] E. S. Ochotta. OADL: A Programming Language for Analog Circuit Synthesis. December, 1989.

[Ochotta 89a] E. S. Ochotta. *The OASYS Virtual Machine: Formalizing the OASYS Analog Synthesis Framework.* Master's thesis, Carnegie Mellon University, 1989.

[Ochotta 92] E. S. Ochotta, R. A. Rutenbar, and L. R. Carley. Equation-Free Synthesis of High-Performance Linear Analog Circuits. *Proc. of Brown/MIT Advanced Research in VLSI and Parallel Systems*, pages 129-143. Providence, RI, March, 1992.

[Ochotta 94] E. S. Ochotta. *Synthesis of High-Performance Analog Cells in ASTRX/OBLX.* Ph.D. thesis, Carnegie Mellon University, 1994.

[Ochotta 94a] E. S. Ochotta, L. R. Carley, and R. A. Rutenbar. Analog Circuit Synthesis for Large, Realistic Cells: Designing a Pipelined A/D Convert with ASTRX/OBLX. *IEEE Custom Integrated Circuit Conference*, pages 15.4/1-4. May, 1994.

[Ochotta 94b] E. S. Ochotta, R. A. Rutenbar, and L. R. Carley. ASTRX/OBLX: Tools for Rapid Synthesis of High-Performance Analog Circuits. *ACM/IEEE Design Automation Conference*. June, 1994.

[Ochotta 96] E. S. Ochotta, R. A. Rutenbar, and L. R. Carley. Synthesis of High-Performance Analog Circuits in ASTRX/OBLX. *IEEE Transactions on Computer-Aided Design.* vol 15, pp 273-294, March, 1996.

[Onodera 90] H. Onodera, H. Kanbara, and K. Tamaru. Operational Amplifier Compilation with Performance Optimization. *IEEE Journal of Solid State Circuits*, vol. 25, pp 466-473, Apr. 1990.

[Parsa 83] H. Parsa and A. L. Tits. Nonuniform, Dynamically Adapted Discretization For Functional Constraints in Engineering Design Problems. *Proc. 22nd IEEE Conf. Decision and Control*, pages 410-411. December, 1983.

[Pelgrom 89] M. J. M. Pelgrom, A. C. J. Duinmaijer, and A. P. G. Welbers. Matching Properties of MOS Transistors. *IEEE Journal of Solid-State Circuits.* SC-24(5):1433-1439, October, 1989.

[Pickerrell 83] T. W. Pickerrell. New Analog Capabilities on Semi-Custom CMOS. *IEEE Custom Integrated Circuit Conference.* 1983.

[Pillage 90] L. T. Pillage and R. A. Rohrer. Asymptotic Waveform Evaluation for Timing Analysis. *IEEE Transactions on Computer-Aided Design.* CAD-9(4):352-366, April, 1990.

[Pletersek 85] Pletersek, T., J, Trontelj, L. Trontelj, I. Jones, G, Sherton, Y. Sun. Analog LSI Design with CMOS Standard Cells. *IEEE Custom Integrated Circuit Conference.* 1985.

[Pletersek 86] T. Pletersek, J. Trontelj, L. Trontelj, I. Jones, and G. Shenton. High-Performance Designs with CMOS Analog Standard Cells. *IEEE Journal of Solid-State Circuits.* sc-21(2):215-222, April, 1986.

[Polak 87] E. Polak. On Mathematical Foundations of Non-Differentiable Optimization in Engineering Design. *SIAM Review.* 2921-89, 1987.

[Powell 72] M. J. D. Powell. *Unconstrained Minimization Algorithms Without Computation of Derivatives.* Technical Report TP483, AERE Harwell Report, 1972.

[Powell 75] M. J. D. Powell. A View of Minimization Algorithms That Do Not Require Derivatives. *ACM Trans. Math Software.* 197-107, 1975.

[Press 88] W. Press, B. Flannery, et. al. Direction Set (Powell's) Method in Multidimensions. *Numerical Recipes in C.* Cambridge University Press, 1988, pages 309-317, Chapter 10.

[Raghavan 93] V. Raghavan, R. A. Rohrer, M. M. Alabeyi, J. E. Bracken, J. Y. Lee, and L. T. Pillage. AWE Inspired. *IEEE Custom Integrated Circuit Conference*, pages 18.1/1-8. San Diego, May, 1993.

[Rhyne 89] G. W. Rhyne and M. B. Steer. Comments on 'Simulation of Nonlinear Circuits in the Frequency Domain. *IEEE Transactions on Computer-Aided Design.* 8(8), August, 1989.

[Richter 83] S. L. Richter and R. A. DeCarlo. Continuation Methods: Theory and Applications. *IEEE Transactions on Circuits and Systems.* CAS-30347-352, June, 1983.

[Rohrer 67] R. A. Rohrer. Fully Automated Network Design by Digital Computer, Preliminary Considerations. *Proceedings of the IEEE.* 55(12):1929-1939, December, 1967.

[Rohrer 92] R. A. Rohrer. Circuit Simulation - The Early Years. *IEEE Circuits and Devices.* 8(3):32-37, May, 1992.

[Romeo 91] F. Romeo and A. Sangiovanni-Vincentelli. A Theoretical Framework for Simulated Annealing. *Algorithmica.* 6302-345, 1991.

[Roylance 75] Roylance, Gerald Lafael. Anthropomorphic Circuit Analysis. Bachelor's thesis, Massachusetts Institute of Technology. 1975 .

[Roylance 80] Roylance, Gerald Lafael. *A Simple Model of Circuit Design.* Master's thesis, Massachusetts Institute of Technology, 1980.

[Rutenbar 89] R. A. Rutenbar. Simulated Annealing Algorithms: An Overview. *IEEE Circuits And Devices Magazine.* 5(1):19-26, January, 1989.

[Rutenbar 93] R. A. Rutenbar. Analog Design Automation: Where are we? Where are we going?. *Proc. IEEE Custom Integrated Circuits Conf.* 13.1.1-13.1.8, May, 1993.

[Rutenbar 95] R. A. Rutenbar, L. R. Carley, P. C. Maulik, E. S. Ochotta, T. Mukherjee, B. Basaran, S. Mitra, S. K. Nag, B. R. Stanisic. Synthesis and Layout for Analog and Mixed Signal ICs in the ACACIA System. *Proc. Advances in Analog Circuit Design Workshop.* Villach, Austria, April, 1995.

[Rychener 85] Rychener. Michael D. Expert Systems for Engineering Design. *Expert Systems.* 2(1), January, 1985.

[Saleh 90] R. A. Saleh and A. R. Newton. *Mixed-Mode Simulation.* Kluwer Academic Publishers, 1990.

[Saleh 94] R. A. Saleh, D. L. Rhodes, E. Christen, and B. A. A. Antao. Analog Hardware Description Languages. *IEEE Custom Integrated Circuit Conference*, pages 349-356. May, 1994.

[Schank 77] Schank Roger C. and Abelson Robert P. *Scripts, Plans, Goals, and Understanding.* Lawrence Erlbaum, Hillsdate N. J., 1977.

[Schoen 93] F. Schoen. On a New Stochastic Global Optimization Algorithm Based on Censored Observations. *Journal of Global Optimization.* 4(1):17-36, January 1993.

[Sechen 87] C. Sechen and K. Lee. An Improved Simulated Annealing Algorithm for Row-Based Placement. *Proceedings of the IEEE/ACM International Conference on Computer-Aided Design*, pages 478-481. 1987.

[Sechen 85] C. Sechen and A. Sangiovanni-Vincentelli. The TimberWolf Placement and Routing Package. *IEEE Journal of Solid-State Circuits.* SC-20(2):510-522, April, 1985.

[Sechen 86] C. Sechen and A. Sangiovanni-Vincentelli. TimberWolf 3.2: A New Standard Cell Placement and Global Routing Package. *Proceedings of the IEEE/ACM Design Automation Conference*, pages 432-439. 1986.

[Seda 92] S. Seda, M. Degrauwe, and W. Fichtner. Lazy-Expansion Symbolic Expression Approximation in SYNAP. *Proceedings of the IEEE/ACM International Conference on Computer-Aided Design*, pages 310-317. Santa Clara, November, 1992.

[Seda 88] S. Seda, M. Degrauwe, and W. Fichtner. A symbolic Analysis Tool for Analog Circuit Design Automation. *Proceedings of the IEEE/ACM International Conference on Computer-Aided Design*, pages 488-491. Santa Clara, November, 1988.

[Serhan 85] Serhan, George I. Automated Design of Analog LSI. *IEEE Custom Integrated Circuit Conference.* 1985.

[Sheu 88] Sheu Bing J., Fung Antony H. and Lai Ying-Nan. A Knowledge-Based Approach to Analog IC Design. *IEEE Transactions on Circuits and Systems.* 35(2):256-258, February, 1988.

[Sheu 87] B. J. Sheu, J. H. Shieh, and M. Patil. BSIM: Berkeley Short-Channel IGFET Model For MOS Transistors. *IEEE Journal Of Solid-State Circuits.* SC-22(4):558-566, August, 87.

[Shyu 88] J-M. Shyu and A. Sangiovanni-Vincentelli. ECSTACY:A New Environment for IC Design Optimization. *Proceedings of IEEE/ACM International Conference on Computer-Aided Design*, pages 484-487. November, 1988.

[Siemens 95] Siemens. Private Communication. . 1995 .

[Stanisic 91] B. R. Stanisic and M. W. Brown. Behavior Modelling of Mixed Analog Digital Circuits. *Applications of VHDL to Circuit Design.* In R. E. Harr and A. G. Stanculescu, Kluwer Academic Press, 1991, Chapter 3.

[Stanisic 93] B. R. Stanisic, R. A. Rutenbar and L. R. Carley. Power Distribution Synthesis for Analog and Mixed-Signal ASICs in RAIL. *Proc. IEEE Custom Integrated Circuits Conf.* 17.4.1-17.4.5, May, 1993.

[Stanisic 96] B. R. Stanisic, R. A. Rutenbar, and L. R. Carley. *Synthesis of Power Distribution to Manage Signal Integrity in Mixed-Signal ICs.* Kluwer Academic Publishers, 1996.

[Stone 84] D. C. Stone, J. E. Schroeder, et. al. Analog CMOS Building Blocks for Custom and Semicustom Applications. *IEEE Journal of Solid-State Circuits.* SC-19(1), February, 1984.

[Strenski 91] P. N. Strenski and S. Kirkpatrick. Analysis of Finite Length Annealing Schedules. *Algorithmica.* 6346-366, 1991.

[Sussman 75] G. J. Sussman, R. M. Stallman. Heuristic Techniques in Computer-Aided Circuit Analysis. *IEEE Transactions on Circuits and Systems.* CAS-22(11), November, 1975.

[Suyama 89] K. Suyama. *Analysis, Simulation and Application of Linear and Non-Linear Switched-Capacitor and Mixed Switched-Capacitor/Digital Networks.* Ph.D. thesis, Columbia University, 1989.

[Swartz 90] W. Swartz and C. Sechen. New Algorithms for the Placement and Routing of Macrocells. *Proceedings of IEEE/ACM International Conference on Computer-Aided Design*, pages 336-339. November, 1990.

[Swings 91] K. Swings, S. Donnay, W. Sansen. HECTOR: A Hierarchical Topology-Construction Program for Analog Circuits Based on a Declarative Approach to Circuit Modelling. *IEEE Custom Integrated Circuit Conference*, pages 5.3/1-5. 1991.

[Swings 91a] K. Swings, G. Gielen, W. Sansen. An Intelligent Analog IC Design System Based on Manipulation of Design Equations. *IEEE Custom Integrated Circuit Conference*, pages 8.6/1-5. 1991.

[Szu 87] H. Szu. Fast Simulated Annealing. *Physics Letters.* 122(3,4):157-162, June, 1987.

[Temes 67] G. C. Temes and D. A. Calahan. Computer-Aided Network Optimization -- The State-of-the-art. *Proceedings of the IEEE.* 55(12):1864, December, 1967.

[Tham 95] K.-M. Tham and K. Nagaraj. A Low Supply Voltage High PSRR Voltage Reference in CMOS Process. *IEEE Journal of Solid-State Circuits.* SC-30586-590, May, 1995.

[Toumazou 90] C. Toumazou, C. A. Makris, and C. M. Berrah. ISAID - A Methodology for Automated Analog IC Design. *IEEE International Symposium on Circuits and Systems*, pages 531-533. IEEE, 1990.

[Trajkovic 91] L. Trajkovic, R. Melville, and S. Fang. Improving DC Convergence in a Circuit Simulator using a Homotopy Method. *Proceedings of IEEE/ACM Custom Integrated Circuit Conference*, pages 8.1.1-8.1.4. May, 1991.

[Trontelj 90] J. Trontelj, L. Trontelj, A. Pletersek, A. Vodopivec, and G. Shenton. Rule Based CAD Tools for Analog Circuit Synthesis and Layout Compilation. *IEEE Custom Integrated Circuit Conference*, pages 14.8/1-4. IEEE, Boston, May, 1990.

[Vanderbilt 84] D. Vanderbilt and G. Louie. A Monte Carlo Simulated Annealing Approach to Optimization over Continuous Variables. *J. Comput. Phys.* 36259-271, 1984.

[Visweswariah 91] C. Visweswariah and R. A. Rohrer. Piecewise Approximate Circuit Simulation. *IEEE Transactions on Computer-Aided Design,* pages 861-870, July 1991.

[Vlach 94] J. Vlach and K. Singal. *Computer Methods for Circuit Analysis and Design.* Van Nostrand Reinhold Company, 2nd edition, 1994.

[Vladimirescu 80] A. Vladimirescu and S. Liu. The Simulation Of MOS Integrated Circuits Using Spice2. *Memo No. UCB/ERL-M80/7.* February, 1980.

[Wambacq 91] P. Wambacq, J. Vanthienen, G. Gielen, W. Sansen. A Design Tool for Weakly Nonlinear Analog Integrated Circuits with Multiple Inputs (Mixers, Multipliers). *IEEE Custom Integrated Circuit Conference*, pages 5.1/1-5. 1991.

[Wambacq 95] P. Wambacq, F. V. Fernandez, G. Gielen, W. Sansen, and A. Rodriquez-Vasquez. Efficient Symbolic Computation of Approximated Small-Signal Characteristics of Analog Integrated Circuits. *IEEE Journal of Solid-State Circuits.* SC-30(3):327-330, March, 1995.

[Williams 84] Williams, Brian Charles. *Qualitative Analysis of MOS Circuits.* Master's thesis, Massachusetts Institute of Technology, 1984.

[Wirth 74] Nicklaus Wirth. On the Design of Programming Languages. *Proceedings IFIP Congress 74*, pages 386-393. North-Holland, Amsterdam, 1974. In [Horowitz 87].

[Yaghutiel 86] Hormoz Yaghutiel, Alberto Sangiovanni-Vincentelli, and Paul R. Gray. A Methodology for Automated Layout of Switched-Capacitor Filters. *IEEE International Conference On Computer-Aided Design.* 444-447, 1986.

[Zakian 73] V. Zakian and L. Al-Naib. Design of Dynamical and Control Systems by the Method of Inequalities. *Proc. IEE.* 120(11), 1973.

Index

GPSR Compliance

The European Union's (EU) General Product Safety Regulation (GPSR) is a set of rules that requires consumer products to be safe and our obligations to ensure this.

If you have any concerns about our products, you can contact us on ProductSafety@springernature.com

In case Publisher is established outside the EU, the EU authorized representative is:

Springer Nature Customer Service Center GmbH
Europaplatz 3
69115 Heidelberg, Germany

Batch number: 10755284

Printed by Printforce, the Netherlands